D0643488

The IMA Volumes in Mathematics and its Applications

Volume 127

Series Editor
Willard Miller, Jr.

Springer
New York
Berlin
Heidelberg
Barcelona
Hong Kong
London
Milan
Paris
Singapore
Tokyo

Institute for Mathematics and its Applications IMA

The **Institute for Mathematics and its Applications** was established by a grant from the National Science Foundation to the University of Minnesota in 1982. The IMA seeks to encourage the development and study of fresh mathematical concepts and questions of concern to the other sciences by bringing together mathematicians and scientists from diverse fields in an atmosphere that will stimulate discussion and collaboration.

The IMA Volumes are intended to involve the broader scientific community in this process.

Willard Miller, Jr., Professor and Director

* * * * * * * * * * *

IMA ANNUAL PROGRAMS

1982–1983	Statistical and Continuum Approaches to Phase Transition
1983–1984	Mathematical Models for the Economics of Decentralized Resource Allocation
1984–1985	Continuum Physics and Partial Differential Equations
1985–1986	Stochastic Differential Equations and Their Applications
1986–1987	Scientific Computation
1987–1988	Applied Combinatorics
1988–1989	Nonlinear Waves
1989–1990	Dynamical Systems and Their Applications
1990–1991	Phase Transitions and Free Boundaries
1991–1992	Applied Linear Algebra
1992–1993	Control Theory and its Applications
1993–1994	Emerging Applications of Probability
1994–1995	Waves and Scattering
1995–1996	Mathematical Methods in Material Science
1996–1997	Mathematics of High Performance Computing
1997–1998	Emerging Applications of Dynamical Systems
1998–1999	Mathematics in Biology
1999–2000	Reactive Flows and Transport Phenomena
2000–2001	Mathematics in Multimedia
2001–2002	Mathematics in the Geosciences
2002–2003	Optimization
2003–2004	Probability and Statistics in Complex Systems: Genomics, Networks, and Financial Engineering

Continued at the back

Brenda Dietrich   Rakesh V. Vohra

Editors

Mathematics of the Internet

E-Auction and Markets

With 18 Illustrations

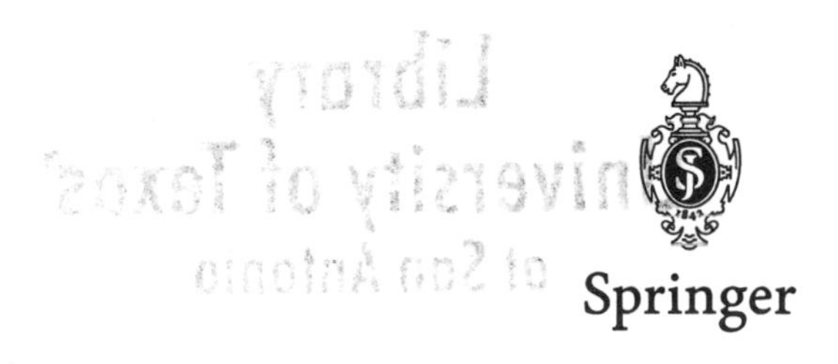

Springer

Brenda Dietrich
Mathematical Sciences
IBM T.J. Watson Research Center
Yorktown Height, NY 10598
USA
dietric@us.ibm.com

Rakesh V. Vohra
Departments of Managerial Economics
and Decision Sciences
Kellogg Graduate School of Management
Northwestern University
Evanston, IL 60208
USA
r-vohra@kellogg.nwu.edu

Series Editor:
Willard Miller, Jr.
Institute for Mathematics and its
Applications
University of Minnesota
Minneapolis, MN 55455, USA

Mathematics Subject Classification (2000): 90-06, 90-08, 90B99, 90C27, 90C90, 9106, 91Bxx

Library of Congress Cataloging-in-Publication Data
Mathematics of the internet : e-auction and markets / editor, Brenda Dietrich, Rakesh V. Vohra.
p. cm — (IMA volumes in mathematics and its applications ; 127)
Includes bibliographical references and index.
ISBN 0-387-95359-0 (alk. paper)
1. Internet auctions. 2. Electronic commerce. 3. Economics, Mathematical. I. Dietrich, Brenda. II. Vohra, Rakesh V. III. IMA volumes in mathematics and its applications ; v. 127.
HF5478 .M38 2001
381′.17′02854678—dc21 2001048434

Printed on acid-free paper.

Production managed by Yong-Soon Hwang; manufacturing supervised by Erica Bresler.
Camera-ready copy provided by the IMA.
Printed and bound by Sheridan Books, Inc., Ann Arbor, MI.
Printed in the United States of America.

9 8 7 6 5 4 3 2 1

ISBN 0-387-95359-0 SPIN 10852174

Springer-Verlag New York Berlin Heidelberg
A member of BertelsmannSpringer Science+Business Media GmbH

FOREWORD

This IMA Volume in Mathematics and its Applications

**MATHEMATICS OF THE INTERNET:
E-AUCTION AND MARKETS**

is based on the proceedings of a very successful "Hot Topics" workshop with the same title. I would like to thank Brenda Dietrich (Department Manager, Mathematical Sciences, IBM T.J. Watson Research Center) and Rakesh V. Vohra (Managerial Economics and Decision Sciences, Kellogg Graduate School of Management, Northwestern University) for their excellent role as organizers of the meeting and for editing the proceedings.

I am also grateful to John Birge (Engineering and Applied Sciences, Northwestern University), Suzhou Huang (Ford Motor Company), and Ennio Stacchetti (Economics, University of Michigan) for serving as co-organizers of the workshop.

Finally, I take this opportunity to thank the National Science Foundation (NSF), Ford Motor Company, IBM, and the University of Minnesota Office of Information Technology, whose financial support of the IMA made this workshop possible.

Willard Miller, Jr., Professor and Director
Institute for Mathematics and its Applications
University of Minnesota
400 Lind Hall, 207 Church St. SE
Minneapolis, MN 55455-0436
612-624-6066, FAX 612-626-7370
miller@ima.umn.edu
World Wide Web: http://www.ima.umn.edu

PREFACE

Inspired by the romance of the internet with auctions, exchanges, dynamic pricing and price search, the Institute for Mathematics and Its Applications organized, in December 2000, a three day workshop focused on just these topics. In spite of Minnesota's well-deserved reputation for rigorous winters, it attracted some 72 participants, from within and beyond the groves of academe as well as from a range of disciplines. Our intent was to bring together researchers from Computer Science, Operations Research, and Economics, and to share industrial experiences with academic experiments. This goal, of cross-community exchange was achieved, through the participation of 20 representatives of industry, and a balanced distribution across the targeted disciplines and mathematics. This volume is a compilation of some of the papers presented at that workshop.

Of the three main topics, auctions is perhaps the most venerable. Auctions have been used since the dawn of civilization to allocate resources and determine prices. The transaction costs associated with running an auction have limited its use. Although some auctions, particularly sealed bid auctions, have traditionally been conducted among physically distributed agents, most auctions, and virtually all small scale auctions, required that the items and the agents be physically co-located for the duration of the auction. Auctions were thus not considered a viable channel for many classes of goods, particularly those that needed to be sold quickly. The internet, with its ability to provide immediate information access and update capability to physically distributed participants, has changed the role of auctions dramatically. Companies like e-Bay encouraged the use of auctions to sell a range of items that were previously sold via posted prices, and gave rise to the concept of peer-to-peer auctions. This in turn generated a trove of data on how buyers and sellers behave in real auctions. Some of this data was at odds with received theory, a delightful circumstance for researchers.

Meanwhile, governments and private companies have been busy trying to sell (or buy) assets and rights whose value was hard to estimate. Auctions seemed like a good way to determine this value. However, for assets such as oil rights and wireless bandwidth, the value of the assets depended on how they were carved up as well as who ended up owning them. This feature, that the value of a combination of items may be less than, or greater than, the sum of the values of the individual items in the combination, posed problems that had not been broadly explored in the theory of auctions. In addition, these combinatorial auctions posed new and difficult problems for both computer scientists and mathematicians. Even if a auction mechanism that exhibits desirable properties can be defined, can the auction mechanism be implemented in a usable, computationally efficient manner?

The urge to privatize electrical power required the privatizers to design a market for the distribution of power which was previously done by monopoly. Advances in information technology and the internet allowed for the monitoring and control of power demand and supply in real time. Again auctions seemed like a good way to set prices. In this case, it was the physics of power generation and distribution that presented new mathematical challenges to auction theory.

Just as the internet reduced the transaction costs of running auctions, it reduced the costs associated with making price changes. Inspired by the airlines' profitable use of yield management, many sellers have begun to experiment with dynamically adjusting their prices in response to changing conditions, and with offering different prices to different groups of customers. Bundled pricing and targeted promotional pricing seems inevitable. In fact it is likely that these pricing strategies will be automated by software agents. Understanding the impact of various strategies on profitability, and characterization of optimal strategies are just two of the questions that immediately arise.

The papers gathered here are a sample of those presented at the workshop to span the areas described above. The reader interested in a full list of the papers presented should consult

http://www.ima.umn.edu/springer/description.html#v127 .

The first paper, by Michael Rothkopf is a compilation of questions about auction design that existing theory does not answer. We hope that this paper will help to focus research in the fields addressed in this meeting.

The next four papers (Dietrich and Forrest, Davenport and Kalagnanam, Eso, Collins and Gini) discuss variations of the winner selection problem in complex auctions. Auctions of assets whose value to the agents depends on how the assets are combined together pose difficult combinatorial optimization problems. Determining the set of winning bids is mathematically challenging, and has been addressed by both search methods and mathematical programming. These papers discuss techniques for solving specific instances of such problems. The paper by Dietrich and Forrest reports on experiments with an integer programming solver that was developed for use by the FCC for the proposed auction #31. Davenport and Kalagnanam describes the winner selection problem in a procurement auction used by a large manufacturer. The paper by Eso proposes an auction for airline seats that would allow buyers to submit bids for a quantity of seats on a specified itinerary and discusses the underlying winner determination problem. Collins and Gini propose a sealed bid auction for selecting bids on coordinated tasks and examine the winner determination problem.

The fifth paper, by Bikhchandani et al, shows how the duality theory of linear programming can be used to design ascending auctions for various combinatorial auctions with nice incentive properties.

The sixth and seventh papers describe the use of auctions to set and adjust prices in a dynamic environment. Both papers are motivated by the problem of capacity utilization in telecommunication. The paper by Massey et al is concerned with how to allocate limited capacity between two different markets. The paper by Anandalingam and Keon focuses on the problem of pricing different classes of telecommunications services.

The last two papers, by Huang, Anderson and Yuang, and Lanning, Massey and Wang address aspects of dynamic pricing. Huang et al is motivated by a class of algorithms that have been proposed for dynamic price adjustment. They formulate and numerically solve a dynamic game between two sellers to understand the structure of the equilibria of the game. A qualitative analysis of the equilibrium suggests limits to the performance of such dynamic price adjustment schemes. Lanning et al propose methods for setting threshold prices for contracts for network capacity in a manner which optimizes the carriers expected revenue.

We hope that this collection of papers will serve as a reference for further research on the mathematics of internet auctions.

Brenda Dietrich
IBM T.J. Watson Research Center

Rakesh V. Vohra
Northwestern University

CONTENTS

below illustrate this. For a fuller treatment, see Rothkopf and Harstad (1994a).

2.1. Should a bidder with more competition bid more aggressively? Let's start with a basic simple question. Should a bidder in a standard sealed bid auction bid more aggressively if she is faced with additional competitors? Auction theory's initial answer to this question was an unambiguous "yes." The answer came from a model in which bidders had independent private values for the object being auctioned. The optimal bid for a bidder in such a model is below her value for the object. To maximize her expected winnings, she must trade off her profit if she wins (which is her value for the object less her bid) against her chance of winning. When her competition increases, her chance of winning with any bid decreases, so she has incentive to increase her bid part of the way towards her value.

For fifteen years, all bidding models were private value models. However, people concerned with construction bidding[5] and, especially, offshore oil lease auctions[6] started considering common value models. In a common value model, the value of what is being auctioned off is uncertain, but what ever it is, it has the same value to each bidder. For example, the value of an offshore oil tract is highly uncertain, but for who ever wins it, the amount of oil in it, the cost of extracting that oil, and the price of oil will be essentially the same. A bidder in a common value situation faced with an increase in competition beyond two competitors should bid *less* aggressively. The reason is that the bidder must compensate for selection bias. In a common value situation, the bidders' estimating errors are the independent random variables. The bidder who is likely to win is the bidder who has made the greatest overestimate of the value.[7] Bidders have to compensate for this effect by bidding less aggressively. The larger the number of bidders, the larger the compensation that is needed. In all of the bidding models I have seen, this need to compensate for selection bias (or the "winner's curse," as is it commonly called) dominates any other effects when a bidder has two or more competitors.

2.2. Why are Vickrey auctions rare? In 1961, William Vickrey proposed and analyzed what he apparently though was a novel auction form.[8] It was a sealed bid auction in which the bidder making the highest bid would win the object being sold, but the price would not be the amount of that bid, but rather the amount of the best losing bid. Such auctions

[5] Rothkopf 1969, See also Wilson 1969.

[6] Capen Clapp, and Cambell 1971.

[7] If bidders with higher estimates bid more, a very reasonable assumption about actual, individually optimal, and Nash equilibrium strategies, then the bidder with the highest estimate will win. Even if all of the estimates are unbiased, the estimate of the winning bidder, the only one that matters economically, will be highly biased.

[8] Vickrey 1961. In fact, the auction form was unusual rather than new. See Lucking-Reiley 2000.

are called "second-price" or Vickrey auctions. It is easy to show that in a single, isolated Vickrey auction it is a dominant (not merely equilibrium) strategy for a bidder to bid exactly what the object being sold is worth to her.[9] If all bidders follow this dominant strategy, the auction result will be perfectly efficient in that the bidder with the highest value will always win. In addition, Vickrey proved a "revenue neutrality theorem." This theorem showed that (in the independent private values context of the day), the average revenue received by the seller would be the same as it would in standard sealed bid auction or a progressive oral auction.

Given these results, Vickrey auctions appear to be a superior auction form, but to this day they remain unusual. Why? The answer is that two modeling assumptions are too strong. One of these assumptions is that it is sufficient to analyze an isolated auction. The other assumption is that bidders will believe that the rules of the auction will always be followed. Rothkopf, Teisberg and Kahn 1990 modeled a Vickrey auction in a broader context. They assumed that the winning bidder would have to negotiate with third parties and that these third parties would be able to capture some fraction of the "economic rent" revealed by the auction. For example, if in an auction with publicly opened bids, the winning bidder offered $ 5 million but only had to pay $ 3 million, everyone she subsequently had to negotiate with would know that she had $ 2 million that she didn't "need." Because of this, they would be able to get from her some extra fraction of that $ 2 million. It turns out that if the logic of Myerson's revenue equivalence theorem[10] can be applied to situations like this with third party rent capture. It shows that what is equivalent is not the bid taker's expected revenue, but the expected amount paid by the bidders. Hence, on average all of the extra money captured by third parties comes out of the pocket of the bid taker.[11]

In addition to this problem from the bid taker's point of view, bidders are wary of Vickrey auctions because they worry about whether the bid taker will follow the rules. If the bid taker in the example above is not completely trustworthy, he may concoct an imaginary or insincere bid of, say, $ 4 million, thus increasing his revenue by $ 1 million. Rothkopf and Harstad 1995 developed two models of bid taker cheating in Vickrey auctions. In both, even a small probability of cheating by bid takers makes sustained, honest Vickrey auctions impossible. Recently, Lucking-Reiley 2000 documented cheating in a Vickrey auction.

[9]No matter what other bidders do, if she loses because she bids less than her value or wins because she bids more than her value, she is worse off.

[10]Myerson 1981 presents a more general version of Vickrey's original revenue equivalence theorem.

[11]In reaction to this analysis, Nurmi and Salomaa 1993 published a cryptographic protocol for Vickrey auctions that limits who knows what about bids. This may deal with this concern about Vickrey auctions, but it does not deal with bidders' concern about cheating by bid takers.

2.3. Are sealed bid auctions efficient? Generally, economists do not believe that standard sealed bid auctions are perfectly efficient. They are aware that occasionally the bidder with the highest value will get too greedy and lose the auction to a bidder with a lower value. Ironically, however, *a priori* symmetric sealed bid auctions are perfectly efficient in the economists' standard models of them.[12] The reason is that in these models the equilibrium bidding strategies are monotonic in the bidder's private estimate of the value to her of what is being sold. Recently, Harstad, Rothkopf and Waehrer[13] showed that this unrealistic result is an artifact of the assumption that a bidder's private information is a scalar. When bidders have a vector of private information, say an estimate of the value to themselves and estimates of the value to each other bidder, the monotonicity of bids in own value estimates disappears.

2.4. Are there n bidders or N potential bidders? Most game theoretic models of auctions, and all of the early ones, assumed that there were a known number, n, of bidders. These models were used to examine the relative revenue to be obtained from different auction forms. See, for example, Milgrom and Weber 1982 who explored this issue using the concept of affiliated values.[14] They found that on average revenue from progressive auctions was greater than or equal to the expected revenue of a Vickrey auction, which in turn was greater than or equal to the expected revenue from standard sealed bidding or a Dutch auction. In addition, in their model having more bidders was better for the bid taker.

However, these conclusions are not necessarily true about real auctions. If instead of assuming that there are n bidders, we assume that there is a pool of potential bidders who must decide whether to incur the cost of participating in the auction, two things may happen. First, the auction form producing the greatest expected revenue can change. Second, the bid taker generally prefers a limited number of bidders to a large number. Implicitly, he is paying the cost of auction participation by the bidders.[15]

2.5. If a bidder offers her maximum bid, does she care if it wins? Until recently, auction theory assumed that bidders were indifferent between winning and losing an item at the maximum bid they were willing to make. With perfect capital markets, that might indeed be the case. However, a few years ago, I consulted for a bidder who valued a license he was bidding on at about $ 85 million dollars. However, he was only able to arrange financing for the license that would allow him to bid $ 65 million.

---

[12]See, for example, the most cited model of Milgrom and Weber 1982 or any of the other models discussed in McAfee and McMillan's 1987 survey paper.

[13]Harstad, Rothkopf and Waehrer, 1996.

[14]Bidders with (strictly) affiliated values will raise their estimate of the value of what is being sold if they find out that another bidder's value is higher. Both independent private values and common values are limiting cases of affiliated values.

[15]See Harstad 1990, Harstad 1993.

He cared a lot about whether his maximum bid of \$65 million won. In this auction, which had significant bid increments, he wanted to be sure to be the first to bid \$65 million.[16] Recently, the auction literature has been augmented by models that consider bidders facing financing constraints.[17]

2.6. Does a bidder's bid affect the future bids of her competitors? The vast bulk of the literature on auction theory deals with models of single, isolated auctions. However, auctions that are an important part of commerce are, almost by definition, not single and isolated. An auction theory that deals only with single, isolated auctions makes no more sense than would chemistry if dealt only with single atoms and not with molecules.

This was brought home to me in the 1960s when my then employer, Shell, asked me to evaluate an analysis of bidding strategy. The situation was that Shell sold a solvent it manufactured, methyl ethyl ketone (MEK), by standard sealed bid to government agencies several times each year. Shell's analyst had done some careful statistical work to back out freight costs and quantity effects. After correcting for these effects, he obtained a narrow probability distribution for the best competitive bid Shell had faced. His analysis, which used this distribution in a decision theory model of a single isolated auction,[18] suggested that Shell could make much more profit from these auctions if it would bid a little more aggressively. The reason that it seemed so much more profitable to bid a bit more aggressively is that, with the narrow distribution of the best competitive bid, doing so would apparently raise Shell's probability of winning from about 50% to about 99.5% at little cost in profit if it won. The problem with this analysis is that there was only one other manufacturer of MEK, Esso (now ExxonMobile). It was unlikely that Esso would let Shell win 99.5% of the government business. Rather, Esso would probably continue to win about half of the auctions at any reasonable price. Hence, the only long run effect of bidding a bit more aggressively would be to lower the price.

A few years later, Shmuel Oren and I wrote a paper in which we considered the effect of a bidder's bid in one auction on the bids of its competitors in future auctions.[19] Rather than solve a sequential game, we used control theory and reaction functions, a construct from nineteenth century economics. In some ways, it would have been nice if we had solved a full sequential game model in which every bidder's dynamic strategy is

---

[16] Even though most progressive auctions have strictly controlled bid increments and no progressive auction allows infinitesimal bid increments, the bulk of auction theory assumes that bids can vary continuously. For a discussion of the effect of finite bid increments, see Rothkopf and Harstad 1994b.

[17] See particularly Che and Gale 1998.

[18] He was using Friedman's (1956) approach and following suggestions in a number of academic and trade papers of the day that discussed using historical data on auctions to estimate the probability distribution of the best competitive bid for use in it.

[19] Oren and Rothkopf 1975.

optimal with respect to every other bidder's dynamic strategy. However, not only would that have been hard mathematically, it is not clear that it would have been behaviorally realistic. Such a model would depend upon what bidders believe their competitors would do in situations that have never occurred. When the same bidders interact regularly, sociology may have as much to say as game theory about how they behave.[20]

3. Some modeling opportunities. I hope that these examples have brought home the idea that modeling and context matter. Next, I want to discuss some of the mathematically difficult issues that arise in attempting to build useful models of real auctions. In the examples above, I have already alluded to several issues such as non-scalar information, endogenous bidder entry determination, capital limitations of bidders, and the strategic effect of sequential auctions. These are all good issues and, while there has been some recognition of them, there is still much to do to integrate them fully them into auction theory. Among them, I think that non-scalar information and sequential auctions will prove the hardest to deal with. Both appear to add complications that are not easily simplified while maintaining some generality.

There are other issues I will discuss as well, and then I want to address two major auction design problems: spectrum auctions and electricity and transmission rights auctions.

3.1. Asymmetry. In most game theoretic auction models, a bidder's strategy is a general function of her private information. The Nash equilibrium that is considered the solution to such game models requires that each bidder's strategy function be optimal with respect to the functions chosen by all other bidders. Finding optimal functions is difficult enough. Finding an equilibrium set of functions is even harder. Often, the only solutions that game theorists have been able to find have been for the special case in which all of the bidders are identical *a priori* and use the same strategy function.[21] This is useful theory, but there are many situations in which there are commonly known asymmetries. Evaluating how different auction rules favor bidders with higher values or the effect of subsidizing high-cost bidders requires asymmetric models.[22] In my view, *a priori* asymmetry is sufficiently difficult to deal with in generality that exploring various approximate methods and limited special cases is useful.

3.2. Auctioning items with interrelated values. When multiple items are auctioned, things can get complicated. In general, we have to think about situations in which a bidder's value for an item can depend

---

[20] See, for example, Smith 1990.

[21] There are a few exceptions. See for example McAfee and McMillan 1989, Maskin and Reiley forthcoming. In addition, Rothkopf 1969 deals with asymmetry in a limited strategy game model and Klemperer 1998 has dealt with small asymmetries.

[22] Rothkopf, Harstad and Fu 1996 have used Rothkopf's 1969 limited-strategy model to evaluate the surprising effects of subsidizing high-cost bidders.

upon which other items she wins.[23] In some cases, items may be truly independent. In some, the dependency may take on a particularly simple form. For example, a bidder may have a value for only one item. Similarly, a bidder may have a budget constraint that limits its ability to buy many items. In some cases, however, we may be tempted to apply auctions to the sale of highly complicated assets. For example, we may wish to auction off the right to use various segments of a region's railroad tracks during particular time segments. Someone who wants to buy the right to send a train from city A to city B during a particular time period needs to collect a complicated pattern of rights. There may be alternative paths and time periods that meet her need, but she has use for only one workable collection. There are also intermediate cases in which different items have synergistic values, but there are no absolute requirements for certain combinations.[24]

In any event, a seller offering multiple items with interrelated values must decide whether to sell them sequentially or simultaneously. If they are to be sold sequentially, in what order will they be sold? If they are to be sold simultaneously, will it be an iterative process or a one-time sealed bid? Will bidders be able to submit bids on combinations of items, e.g., a bid on A and B? Will bidders be able to submit contingent bids, e.g., a bid on A if I don't win B; budget constrained bids, e.g., all other bids are withdrawn once I have spent $ 1,000; or OR-bids, e.g., $ X for A or $ Y for B, but not both? Whatever the form of the sale, bidders need to be able to figure out their chances of acquiring synergistic combinations or of staying within their budgets. Will the allocation of items to bidders tend to be efficient? Will the auction allow bidders wanting multiple items to collude (tacitly or explicitly) in order to avoid competing and driving up prices? Where bidders have private information that is of use to other bidders, will the auction form encourage revelation or will it produce incentives for bidders to keep the information private?[25]

Both the FCC spectrum license auctions and electricity auctions discussed below involve multiple items with interrelated values.

---

[23]Also of potential significance and of great difficulty, but not discussed further here, is the situation in which a bidder's value for an item depends upon who wins some other item. For example, if there are two broadcasting licenses for a region being sold, my value for a license might depend upon whether a particularly tough competitor wins the other license. Jehiel, Moldovanu and Stacchetti 1999 discuss a related but simpler situation in which there is a single item but bidders care who wins it if they lose.

[24]No discussion of bidding on combinations of items would be complete without pointing out that the definitions of items are not preordained. Rather, the seller defines them. That definition may be critical. Should a seller sell each one-bedroom condo in a building separately, or should he pool them and sell the right to choose one of the unsold condos from the pool? Should the Federal government sell coal leases or a one-year option to sign such a lease? The answer to this question, discussed in Rothkopf and Englebrecht-Wiggans 1992, could be worth many millions of dollars to the US treasury. Should the FCC sell three 10 MHz licenses in a region, two 15 MHz licenses, or one 30 MHz License?

[25]See Hausch 1986.

3.3. Transaction costs. Most auction theory has been developed on the assumption that transaction costs are negligible. Even when the items being sold are highly valuable, that assumption is suspect, as bidders will often spend a few percent of the value at stake to evaluate the items and participate in the auction. When the items are inexpensive, the assumption can be quite misleading. Just the time of the auction participants can be significant. The daily produce auctions held in Vineland, New Jersey are simple progressive auctions that, on average, take over 20 seconds per transaction. Many transactions involve less than $ 100 worth of produce. Typically, there are 40 to 50 buyers in the auction room, several staffers supporting the auctioneer, and up to 100 farmers queued up for the chance to sell their produce. In order to reduce the time required to complete a transaction, the auctioneer delivers the sale document to the winning bidder by throwing it tucked into a slit cut in an old tennis ball. By contrast, the Dutch auctions used in the Dutch flower markets average four seconds per transaction.[26]

The role of transaction costs cries out for analytic attention. Arguably, the Internet has lowered the cost of holding an auction by an order of magnitude, making auctions relatively more attractive compared to posted prices and negotiations. This needs to be studied and modeled as well as being checked empirically.[27]

Transaction costs can affect the relative attractiveness of different auction forms. Recently, Lucking-Reiley 1999 found to his surprise that in a controlled experiment with real Internet auctions, Dutch auctions produced significantly more revenue than did standard sealed bidding. The strategic and revenue equivalence of Dutch auctions and standard sealed bidding has been unchallenged in the theoretical literature. In each, selecting a bid involves exactly the same trade off between the probability of winning and the profit from winning, and in each the choice must be made before observing any competitive bid. I suspect that Lucking-Reiley's result subtlety involves transaction costs. Lucking-Reiley's Internet Dutch auctions are "slow Dutch auctions," more like Filene's basement where unsold goods are periodically reduced in price than the four-second Dutch flower auctions. If I see something in Filene's basement I think will be attractively priced in two weeks, I can buy it now or come back in two weeks when the price is lower. However, if I wait, I not only risk that the item will be sold, but I also have to pay the cost of returning. This cost can make me decide to pay more now when, if returning were free, I would wait. Lucking-Reiley's Dutch auction bidders had to log back in at a later time in order to bid, and the magnitude of the revenue difference is about that of the cost of doing so. I doubt that this is an isolated example. Online

[26]Kambil and van Heck 1998 discuss the Dutch flower auctions. The Vineland information is from personal observation.

[27]Kambil and van Heck 1998 provide an empirically based framework for analyzing regular auction exchanges but no mathematical models.

auctions provide a whole new context for auctions. Old assumptions about auction form will have to be reexamined in it. This will surely require good observation and modeling. I am unsure whether the mathematics will be particularly difficult.

3.4. Combinatorial spectrum rights auctions. The FCC's spectrum rights auctions were a wake up call for auction theorists. Several companies filed briefs with the FCC (FCC Docket 93-253) accompanied by papers by economists suggesting alternative rules for the spectrum auctions. Five different papers by five sets of leading economists all argued that existing theory implied that there was only one way to hold the auctions, but they proposed five different ways. Much discussion made it clear that existing bidding theory had little to say about situations as complicated as the one faced by the FCC. What evolved was a series of simultaneous progressive auctions that had their roots in a proposal by Milgrom and Wilson. The history of these auctions is discussed extensively elsewhere.[28]

One aspect of the auction design, however, in the view of Congress and the FCC, needed additional attention. The simultaneous progressive auction design did not allow bidders to place bids on combinations of licenses.[29] However, the FCC has now announced an auction design for a simultaneous progressive auction with combinatorial bids for a sale in 2001. Although the FCC hosted a conference of theorists to discuss the design of combinatorial auctions, the actual design is quite different from what was discussed. Their auction design and alternatives to it need study by theorists.[30] Among the questions I view as open are

1. Should an auction with combinatorial bids allowed be a one-time sealed bid or a simultaneous progressive auction? What will be the effect on efficiency and revenue?

---

[28]See Cramton 1995, 1997; Cramton and Schwartz 2000; McAfee and McMillan 1996; McMillan 1994 and Bykowsky, Cull and Ledyard 2000.

[29]When the original FCC auctions were designed, the most important argument against allowing combinatorial bids was mathematical. The problem of picking the revenue maximizing set of bids when bids on any combination are allowed is NP complete. Since the FCC sold as many as 1500 licenses in a single sale, this was taken as a reason to disallow bids on any combinations. However, Rothkopf, Pekec and Harstad 1998 point out that winner determination is not NP complete for many economically interesting kinds of combinations. Furthermore, algorithms, including standard integer programming codes like CPLEX, normally perform much better than the worst-case assumption involved in the definition of NP completeness. (This is both my own experience and consistent with Anderson, Tenhuen and Ygge 2000). Thus it may make sense to allow arbitrary combinatorial bids provided that there is a reasonable contingency plan for the unlikely event that winner determination is difficult. One possibility is that bidders prioritize their combinations and that the winner determination calculation start with each bidder's highest priority combination and add additional combinations in priority order until it runs out of calculation time. See Park and Rothkopf 2000. Another proposal discussed in Rothkopf, Pekec and Harstad 1998 is to allow bidders a limited time to challenge a potentially suboptimal winner determination by proposing one that produces more revenue.

[30]Much material is posted on the FCC website http://combin.fcc.gov/papers.html#comments.

2. Should bidders be allowed to offer unrestricted OR-bids, budget based OR-bids, or no OR-bids at all? Does this answer depend upon whether the auction is progressive or a one-time sealed bid?
3. What, in general, should be done if winner determination in not provably optimal?[31]

3.5. Electrical power and transmission rights. The realization that there are no longer economies of scale in electricity generation has led, in many places, to the deregulation of generation and to auctions for the purchase of electricity. Auctions are, or soon will be, involved with the allocation of electricity transmission capacity as well as the provision of "ancillary services" to electricity generation.[32]

It is harder to imagine a more demanding challenge for auction design. The demand for electricity varies substantially by time of day, is uncertain, and in the short run, is extremely inelastic. With unimportant exceptions, electricity cannot be stored, and demand must be met the instant it occurs.[33] However, much of the lowest cost generating equipment takes many hours and incurs substantial costs to start or stop. Furthermore, electricity transmission adds greatly to the challenge. One cannot transmit electricity in a modern grid from A to B without causing flows in essentially every link of the grid. Ignoring line losses, the fact that electricity uses alternating rather than direct current, and the nonlinear constraints associated with reliability concerns, we can construct a linear model to describe the flows. However, all three of these factors are nonlinear and sometimes quite significant. Finally, our electrical system is not a single closed system under a single management, but rather a set of interconnected separate systems. What goes on in the New England system and the Pennsylvania-New Jersey-Maryland (PJM) system affects the New York system but is not under its control.

The problems and disagreements start at the most basic level. What are the commodities to be auctioned? An answer being tried in California is that the basic commodity is electric power during a given hour delivered in a given zone. Twenty-four hours of supply are being procured simultaneously, but separately. This is simple, but it presents several problems. First, generators must deal with the fact that their cost in a given hour may

[31]The 2001 auction has only 12 licenses, so this is not an issue in this auction. It might be for subsequent ones. The issue is more important than the size of the inefficiency that might result. Auctions are used to guarantee fairness. If different suboptimal alternatives favor different bidders, the fairness issue must be addressed.

[32]These services include automatic frequency control, spinning and stand by reserves, and cold start capabilities. They are supplied, typically, by equipment that might otherwise be supplying electricity itself.

[33]One implication of inelastic demand for a nonstorable product is that when demand gets near generating capacity a generator with even a modest amount of capacity can drive the price extremely high by withholding part of it. Indeed, price spikes orders of magnitude higher than normal prices have been observed even when capacity is apparently sufficient to meet demand.

depend heavily upon which other hours it is generating.[34] Second, the appropriate definitions of zones depend upon where the transmission grid is congested, but this depends upon which power is purchased. Attempts to define uncongested zones *a priori* have a spotty record.

Another approach is for bidders to submit bids for both power and start up conditions and costs. The bid taker then solves, exactly or approximately, an integer programming problem to select the bids that meet the projected demand at the lowest cost while respecting the various transmission and operating constraints. Payment is based upon market clearing prices adjusted for locational marginal prices. Variants of this approach are being tried in New York and PJM.

Neither approach is ideal. Evaluations of the approaches have gotten entangled with ideological concerns about the role of the central system operator. Furthermore, each approach is subject to variants in how it is implemented. England is in the process of switching from market clearing prices used in California, NY and PJM to pay-your-bid. In the context of the daily repetition of the auctions, this may not be a bad idea.[35]

There is currently an active debate on the difficult issue of best way to sell transmission rights. Some focus on the many constraints in the network and some want a liquid market in a few popular or representative transmission routes. The problem is complicated by the fact that there are a lot of potential constraints. Even without considering AC issues and reliability contingency constraints, the capacity of every link in the grid is constrained.[36] There are further issues in defining what is to be sold. Should it be an option or a right and duty? The distinction matters because failure to use a right can limit counter flows. An option suggested by O'Neill *et al.* 2000 is to hold an auction that includes option and rights-and-duties on each constraint and on each popular transmission route, including them all in one constrained optimization problem. Such an auction could be held annually, leaving the redistribution of transmission rights to an aftermarket, or daily or even hourly.

4. Conclusions. There are two major approaches to the creation of useful mathematical models of auctions. One is to start with a simple mathematical model and see how much can be solved. It is a common approach, congenial to mathematicians, and it has produced some interesting, beautiful, and occasionally useful mathematics. This has some value. However, it has repeatedly missed important factors that affect how real auctions work, at time misleading practitioners who tried to use the results of the mathe-

[34]An attractive alternative suggested by Elmaghraby and Oren 1999 is to make the commodity a given amount of power starting at time A and running until time B. This can viewed as purchasing horizontal rather than vertical slices of the load-duration curve. It would allow generators to know their start up costs.

[35]See Rothkopf 1999.

[36]See Chow and Peck 1996.

matics for decision making.[37] The alternative approach, which I advocate, is to pay attention to auctions—study their context, purposes, particulars, and peculiarities—and then try to develop mathematical models and theories that, at least approximately, deal with these realities. This is hard and will often lead to less sweeping results. However, the results are much more likely to be useful for answering important questions about real auctions.

As deregulation and lowered transaction costs of the Internet push auctions into new applications and contexts, challenges arise for those trying to develop theories useful for decision making. I have cataloged some factors that need to be incorporated into the analysis of auctions and the mathematical models used in them. These include asymmetry, financially constrained bidders, complicated information structures, bidder decisions about participation, the effects of repeated auctions, auctioning items with interrelated values, and transaction costs. I have also discussed two major areas where new, complicated auctions are being designed—combinatorial spectrum auctions and electricity and transmission rights auctions. In both of these areas, theory to guide decision makers is woefully inadequate. I hope that these challenges will lead the mathematically talented to undertake some work of great potential value.

Acknowledgement and disclaimer. This paper is based upon a presentation to the IMA Workshop on "Mathematics of the Internet: E-Auctions and Markets" on December 3, 2000. Over the years, I have benefited from discussions with bidders, those conducting auctions and most of the smart people who have worked on bidding theory. I have also been supported, in part, by Shell, Xerox, DOE, Rutgers, and NSF grants. However, the idiosyncratic opinions expressed here are solely my own and should not be attributed to any organization or individual with which I have worked or who was kind enough to give me information or financial support.

REFERENCES

ANDERSSON, ARNE, MATTIAS TEHHUNEN, AND FREDRIK YGGE, "Integer Programming for Combinatorial Auctions Winner Determination," International Conference on Multi-Agent Systems, 2000.

BYKOWSKY, MARK M., ROBERT J. CULL AND JOHN O. LEDYARD, "Mutually Destructive Bidding: The FCC Auction Design Problem," *Journal of Regulatory Economics* **17**(3), pp. 205–228, 2000.

CAPEN, EDWARD R., ROBERT CLAPP, AND WILLIAM CAMPBELL, "Biding in High Risk Situations," *J. Petroleum Technology* **23**, pp. 641–53, 1971.

CHAO, HUNG-PO AND STEPHEN PECK, "A Market Mechanism for Electric Power Transmission," *Journal of Regulatory Economics* **10**(1), 1996.

[37] In addition to the economists' briefs to the FCC and Shell's experience with MEK mentioned above, see Capen, Clapp and Campbell 1971.

Che, Yeon-Koo and Ian Gale, "Standard Auctions with Financially Constrained Bidders," *Review of Economic Studies* **65**, pp. 1–21, 1998.

Cramton, Peter, "Money Out of Thin Air: The Nationwide Narrowband PCS Auction," *Journal of Economics and Management Strategy* **4**, pp. 431–495, 1995.

Cramton, Peter, "The FCC Spectrum Auctions: An Early Assessment," *Journal of Economics and Management Strategy* **6**, pp. 267–343, 1997.

Cramton, Peter and Jesse A. Schwartz, "Collusive Bidding: Lessons from the FCC Spectrum Auctions," *Journal of Regulatory Economics* **17**(3), pp. 229–252, 2000.

Elmaghraby, Wedad and Shmuel Oren, "The Efficiency of Multi-Unit Electricity Auctions," *The Energy Journal* **20**(3) pp. 89–116, 1999.

Friedman, Lawrence, "A Competitive Bidding Strategy," Operations *Research* **4**, pp. 104–12, 1956.

Friedman, Lawrence, "Competitive Bidding Strategies," Ph.D. Dissertation, Case Institute of Technology, 1957.

Friedman, Milton, *A Program for Monetary Stability*, Fordham University Press, New York, 1960.

Harstad, Ronald M., "Alternative Common-Value Auction Procedures: Revenue Comparisons with Free Entry," *Journal of Political Economy* **98**, pp. 421–29, 1990.

Harstad, Ronald M. and Michael H. Rothkopf, "An 'Alternating Recognition' Model of English Auctions," *Management Science* **46**, pp. 1–12, 2000.

Harstad, Ronald M., Michael H. Rothkopf, and Keith Waehrer, "Efficiency in Auctions When Bidders Have Private Information About Competitors," *Advances in Applied Micro-Economics*, Volume **6**, Michael Baye, Ed., JAI Press, pp. 1–13, 1996.

Hausch, Donald B., "Multi-Object Auctions: Sequential vs. Simultaneous Sales," *Management Science* **32**, pp. 1599–1610, 1986.

Jehiel, Philippe, Benny Moldovanu, and Ennio Stacchetti, "Multidimensional Mechanism Design for Auctions with Externalities," *Journal of Economic Theory* **85**, pp. 258–293, 1999.

Kambil, Ajit and Eric van Heck, "Reengineering the Dutch Flower Auctions: A Framework for Analyzing Exchange Operations," *Information Systems Research* **9**, pp. 1–19, 1998.

Klemperer, Paul, "Auctions with Almost Common Values: The 'Wallet Game' and its Applications," *European Economic Review* **42**, pp. 757–769, 1998.

Klemperer, Paul, "Auction Theory: A Guide to the Literature," *Journal of Economic Surveys* **13**(3), pp. 227–286, July 1999.

Levin, Dan and James L. Smith, "Equilibrium in Auctions with Entry," *American Economic Review* 84, pp. 585–599.

Lucking-Reiley, David, "Using Field Experiments to Test Equivalence Between Auction Formats: Magic on the Internet," *American Economic Review* **89**, pp. 1063–1080, 1999.

Lucking-Reiley, David, "Vickrey Auctions in Practice: From Nineteenth-Century Philately to Twenty-First Century E-Commerce," *Journal of Economic Perspectives* **14**(3), pp. 183–192, 2000.

Maskin, Eric and John Riley, "Asymmetric Auctions," *Review of Economic Studies*, forthcoming.

McAfee, R. Preston and John McMillan, "Auctions and Bidding," *J. Econ. Literature* **25**, pp. 699–738, 1987.

McAfee, R. Preston and John McMillan, "Government Procurement and International Trade," *Journal of International Economics* **26**, pp. 291–308, 1989.

McMillan, John, "Selling Spectrum Rights," *Journal of Economic Perspectives* **8**, pp. 145–162, 1994.

Milgrom, Paul R. and Robert J. Weber, "A Theory of Auctions and Competitive Bidding," *Econometrica* **50**, pp. 1089–1122, 1982.

Myerson, Roger B., "Optimal Auction Design," *Mathematics of Operations Research* **6**, pp. 58–73, 1981.

NURMI, HANNU AND ARTO SALOMAA, "Cryptographic Protocols for Vickrey Auctions, *Group Decision and Negotiation* **4**, pp. 363–373, 1993.

O'NEILL, RICHARD P., BENJAMIN HOBBS, WILLIAM STEWART, MICHAEL H. ROTHKOPF AND UDI HELMAN, "A Bidder-Specified Transmission Rights Contracts Auction," paper at MEET/EPRI Conference, Stanford, University, August 17, 2000.

OREN, SHMUEL S. AND MICHAEL H. ROTHKOPF, "Optimal Bidding in Sequential Auctions," *Operations Research* **23**, pp. 1080–1090, 1975.

ROTHKOPF, MICHAEL H., "A Model of Rational Competitive Bidding," *Management Science* **15**, pp. 362–373, 1969.

ROTHKOPF, MICHAEL H., "On Auctions with Withdrawable Winning Bids," *Marketing Science* **10**, pp. 40–57, 1991.

ROTHKOPF, MICHAEL H., "Daily Repetition: A Neglected Factor in the Analysis of Electricity Auctions," *The Electricity Journal* **11**, pp. 60–70, April 1999.

ROTHKOPF, MICHAEL H. AND RICHARD ENGELBRECHT-WIGGANS, "Innovative Approaches to Competitive Mineral Leasing," *Resources and Energy* **14**, pp. 233–248, 1992.

ROTHKOPF, MICHAEL H. AND RONALD HARSTAD, "Modeling Competitive Bidding: A Critical Essay," *Management Science* **40**, pp. 364–384, 1994a.

ROTHKOPF, MICHAEL H. AND RONALD M. HARSTAD, "On the Role of Discrete Bid Levels in Oral Auctions," *European Journal of Operational Research* **74**, pp. 572–581, 1994b.

ROTHKOPF, MICHAEL H. AND RONALD M. HARSTAD, "Two Models of Bid-Taker Cheating in Vickrey Auctions," *Journal of Business* **68**, pp. 257–267, 1995.

ROTHKOPF, MICHAEL H., RONALD M. HARSTAD, AND YUHONG FU, "Is Subsidizing Inefficient Bidders Actually Costly?" RUTCOR Research Report # 38–96, Rutgers University, New Brunswick, N.J., 1996, revised 2000.

ROTHKOPF, MICHAEL H., ALEKSANDAR PEKEC, AND RONALD M. HARSTAD, "Computationally Manageable Combinational Auctions," *Management Science* **44**, pp. 1131–1147, 1998.

ROTHKOPF, MICHAEL H., THOMAS J. TEISBERG, AND EDWARD P. KAHN, "Why Are Vickrey Auctions Rare?" *Journal of Political Economy* **98**, pp. 94–109, 1990.

SMITH, CHARLES W., "Auctions: The Social Construction of Value," University of California Press, Berkeley, California, 1990.

VICKREY, WILLIAM, "Counterspeculation, Auctions and Competitive Sealed Tenders," *J. of Finance* **41**, pp. 8–37, 1961.

WILSON, ROBERT B., "Competitive Bidding with Disparate Information," *Management Science* **15**, pp. 446–8, 1969.

A COLUMN GENERATION APPROACH FOR COMBINATORIAL AUCTIONS

BRENDA DIETRICH* AND JOHN J. FORREST*

Abstract. A column generation approach is used to determine the set of winning bids in a combinatorial auction for multiple, unique items. This approach is appropriate for auctions in which an individual agent places bids on multiple combinations, and will accept, subject to some restrictions, one or more of the specified combinations. Preliminary computational results appear.

1. Introduction. Auctions have been used in the sale of items for many years. Auctions provide a means of making scarce items available to a large audience, and of ensuring that the seller receives the maximum revenue for the items. Many forms of auctions exist, including open-cry auctions, silent auctions, ascending bid auctions and descending bid auctions. With the advent of the Internet and other forms of electronic communications, auctions are becoming increasingly common.

Most auctions are for single items, or for multiple copies of the same item. Auctions for related items are often held simultaneously. There are numerous instances where the value to a buyer (or, in our terminology, an agent) of a set of items may not equal the sum of the value (to that agent) of the individual items. In some cases the value of a collection of items, such as an airline ticket, a hotel room, and theater tickets, may be higher than the sum of the values of the individual items. In this case the items (airline ticket, hotel, theater tickets) are said to complement one another. In other cases the sum of the values of the individual items, such as theater tickets and concert tickets for the same time and date, may be higher than the value of the combination. In this case the items are said to substitute for one another. Therefore an agent, attempting to obtain a collection of items or to obtain one collection from a set of specified collections, through participation in several single item auctions, is faced with a dilemma. She may bid on all the items in her desired collection, and obtain some, but not all, of the items in that collection. Even if the total amount she bids does not exceed her value for the collection, she may end up paying more for the items she receives than she values that subset of the collection. Similarly, if she bids for items that substitute for one another, such as the theater tickets and concert tickets, she may end up with both items, at a cost greater than her value for the pair, even if the amount she bids for each item does not exceed her value for that item.

Combinatorial auctions, in which an agent can bid for combinations of items, are beginning to be used. Unfortunately, whereas determining the

*Mathematical Sciences Department, IBM T.J. Watson Research Center, Yorktown Height, NY 10598.

winner of a single item auction is trivial, even when the auction is held over the Internet with thousands of participating agents, determining the winner of a combinatorial auction, even a relatively small combinatorial auction, is far more difficult. For n items, there are $2^n - 1$ possible combinations. Of course, it is not expected that bids will be placed on every possible combination. However, the problem of finding from among a large set of bids, each for a different combination of items, the set of bids that includes each item at most once is known to be computationally difficult. That is, it is a member of a collection of problems called NP-complete, which means that there are no known computational methods that are guaranteed to solve the problem in a number of computations that is bounded by a polynomial in the size of the input data. In practice, exact solution methods for NP complete problems typically require computation resources that grow very rapidly with the problem size, rendering them impractical for all but the smallest instance of the problem. Rothkopf et al. (1998) provides an excellent survey of combinatorial auctions, including a discussion of particular forms of auctions that can be efficiently solved

For combinatorial auctions, where bidding is typically either continuous or in rounds, and each agent requires information about whether each of his bids is "currently" a winning bid, the set of winning bids must be computed quickly, and re-computed as new bids are added or the value bids are increased. We propose a column generation approach for determining the set of winning bids for combinatorial auctions. Our method allows for the use of certain forms of side constraints, and allows for rapid calculation of minimum bid values and the determination of tied solutions. Our formulation also allows for rapid updates as new bids are submitted, and provides a natural means for fast calculation of approximate feasible solutions rapidly. Preliminary computational experience is promising.

Our column generation approach supports two forms of combinatorial bidding, arbitrary conjunctions (ands) and a limited set of disjunctions (ors). Research in the area of combinatorial bid representation remains active. Nisan (1999) provides a lucid discussion of bid representations. We illustrate our column generation approach with a representation chosen specifically to allow an agent to limit his total liability while placing bids on an arbitrary number of arbitrary combinations. Alternate representations can be accommodated by appropriately modifying the column generation procedure.

Each agent is allowed to place an arbitrary number of bids. Each bid can contain any combination of items. A bid consists of the following data elements: agent id, bid value, and items included in the bid. Each player can also specify a set of bid types, and can assign one or more type to each of his bids. Preference for combinations of items is expressed by placing the items in the same bid. Substitution of bids (that is, the desire to have one, but not both, or a pair of bids) is expressed by assigning the bids different types. A winning combination of bids will satisfy the following properties:

(A) each item is included in only one winning bid.

(B) all of the winning bids belonging to each agent must be included in a single type.

Furthermore, among all combinations of bids satisfying these two constraints, the winning combination must maximize total revenue, given as the sum of the values of the winning bids.

To understand property (B), note that we allow a bid to be assigned to more than one type. If a player has a bid $b1$ that is assigned to type $t1$, a bid $b2$ that is assigned to $t1$ and to $t2$, and a bid $b3$ that is assigned to $t2$, then a winning combination can include $b1$ and $b2$, because they both belong to type $t1$. A winning combination could also include $b2$ and b3, because they both belong to type $t2$. The pair $b2$ and $b3$ cannot appear in any winning combination because there is no type that includes both of these bids.

Standard combinatorial auctions as described in Rothkopf et al. (1998) correspond to the case where each player assigns all of his bids the same type. Exclusive-or auctions, where each player is allowed to win only one combination, correspond to the case where each player assigns each bid a unique type.

We note that this specification can easily be converted to the case where each bid is assigned to only one type, through the use of duplicate bids. In the example above, bid $b2$ would be submitted twice by the player, once assigned type $t1$ and once assigned type $t2$. Allowing multiple types to be assigned to a bid allows for a smaller optimization problem, and may allow for more compact user interfaces.

We first present an obvious integer programming formulation of the winning bid selection program. This formulation is an extension, incorporating the type constraint, of the formulation given by Rothkopf et al. (1998).

2. Formulation. Let I denote the set of items, P denote the set of agents, and B denote the set of bids. For each $k \in B$ let $S_k \subseteq I$ denote the set of items in the bid, $v_k \geq 0$ denote the value of the bid, and $p(k) \in P$ denote the agent. We let T denote the set of types, and for each agent $p \in P$ and each type $t \in T$ we let $B_{p,t}$ denote the set of bids of type t made by agent p. Note that, as remarked above, for a pair of types $t \neq t'$, and an agent p the sets $B_{p,t}$ and $B_{p,t'}$ need not be disjoint (that is, a bid made by an agent can be of more than one type), but that for a pair of agents $p' \neq p$ the sets of bids submitted by the agents, $B_{p,t}$ and $B_{p',t'}$ are disjoint. That is, each bid is associated with one or more types, but with only one agent. We note that because of the type constraint, it is not possible to filter the bids so as to include, for each set of items, only a single bid or a single agent.

To designate whether a bid $k \in B$ is included in the winning combination of bids, we use decision variables x_k, each of which must take either the value 0 (indicating that the bid is not in the winning combination) or the value 1 (indicating that the bid is in the winning combination). We also introduce additional 0-1 variables to indicate which type is selected for each agent. For each $p \in P$ and $t \in T$ we let $y_{y,t}$ be 1 if the selected bids of agent p are from type t, and 0 otherwise.

The set of winning bids can be determined by solving the following integer program:

$$Max \sum_{j \in B} v_j x_j.$$

Subject to

$$(1) \qquad \sum_{i \in S_j} x_j \leq 1 \qquad \text{for all} \quad i \in I$$

$$(2) \qquad \sum_{t \in T} y_{p,t} \leq 1 \qquad \text{for all} \quad p \in P$$

$$(3) \qquad x_j - \sum_{j \in B_{p,t}} y_{p,t} \leq 0 \qquad \text{for all} \quad j \in J$$

$$(4) \qquad x_j \in \{0, 1\} \qquad \text{for all} \quad j \in J$$

$$(5) \qquad y_{p,t} \in \{0, 1\} \qquad \text{for all} \quad p \in P, t \in T.$$

Constraint (1) says that each item is in at most one winning bid. If it is required that each item be in exactly one winning bid (that is, that all of the items be sold) then the inequality should be changed to an equality. In this case, if there is an item for which no bids are submitted, the integer program may be infeasible. To prevent infeasibility and to also represent a minimum value for each item, one can include a dummy bid for each individual item, with the value being the reserve value for that item. Constraint (2) says that at most one type is selected for each agent, and constraint (3) says that if a bid is selected, then for that agent a type that includes that bid has also been selected. Constraints (4) and (5) enforce integrality on the decision variables. If each agent uses only one type of bids, then the y variables for that agent are not required and constraints (2), (3) and (5) are deleted. Similarly, if for agent p each type includes at most one bid and each bid has a unique type, then the y variables for agent p are not required. We use B_p to designate the bids from agent p, delete constraint (3) for agent p and replace constraint (2) for agent p by the following, simpler constraint:

$$(2') \qquad \sum_{j \in B_p} x_j \leq 1.$$

In the case where each bid is assigned to one type, there will be a single y variable in constraint (3).

The integer program (1)–(5) is easy to formulate from the auction data. Any integer programming solver, such as OSL or CPLEX (cf. Andersson et al. (1999)) can be used to solve problems of this form.

However, the computation time and the amount of computer memory required to solve even moderate sized auctions with this formulation may be excessive, as the number of items or bids increases.

We use a column generation formulation that, for small to moderate sized problems, appears to require remarkably little computation time and computer memory to solve the problem. We expect that these computational advantages will carry over to some larger problems, where it may not be necessary to specify the full optimization problem in order to find the winning combination. Furthermore, this approach permits a wide variety of side constraints on permissible combinations of bids from any one agent.

Rather than consider each bid individually, and use a number of constraints to indicate which combinations of bids can be selected simultaneously, we generate, for each agent, the combinations of that agent's bids that can be simultaneously accepted. We call such combinations proposals, because each represents a "proposed" set of bids to be awarded to the agent. Requirement (A) says that the bids belonging to a proposal can have no items in common. Requirement (B) says that the bids belonging to a proposal must all be included in a single type.

Various methods can be used to generate the set of proposals. If $B_{p,t}$ contains only a few bids, an enumeration and elimination process is very fast. For each $p \in P$ and $t \in T$, enumerate all subsets of $B_{p,t}$. From this collection of sets of bids, eliminate all sets C such that there exists $j, j' \in C$ with $S_j \cap S_{j'} \neq \phi$. Denote the collection of remaining sets of bid by $\Phi_{p,t}$. When $B_{p,t}$ contains many bids the following algorithm can be used to construct $\Phi_{p,t}$. We may assume that the bids in $B_{p,t}$ are labeled $1, 2, 3, ..., n$.. For each positive integer $k = 1, 2, 3, ..., n$ let $\Phi^k_{p,t}$ denote the collection of subsets of $\{1, 2, 3, ..., k\}$ such that for each$j, j' \in C \in \Phi, S_j \cap S_{j'} \neq \phi$. Note that $\Phi^1_{p,t} = \{\phi, \{1\}\}$ is trivial to construct, and note that the set $\Phi^k_{p,t} = \Phi^{k-1}_{p,t} \cup \{C \cup \{k\} \; s.t. \; C \in \Phi^{k-1}_{p,t} \; and \; \bar{S}_C \cap S_k = \phi\}$ can be constructed from $\Phi^{k-1}_{p,t}$ by copying $\Phi^{k-1}_{p,t}$ and determining whether the bid k can be added to each element of $\Phi^{k-1}_{p,t}$.

For each $C \in \Phi_{p,t}$ we use $\overline{S}_C$ to denote the set of items in the bids in proposal C, that is, $\overline{S}_C = \cup_{k \in C} S_k$. The amount that player p would pay for the proposal C is denoted w_C and is equal to the sum of the value of the bids in C, given by $w_C = \sum_{j \in C} v_j$. Once the sets $\Phi_{p,t}$ are computed, the winning combination of bids can be determined by selecting at most one proposal for each agent, while requiring that each item is in at most one selected combination.

We let $\Phi_p = \cup_t \Phi_{p,t}$ be the proposals for agent p and we let $\Phi = \cup_p \Phi_p$ be the set of all proposals. For each $C \in \Phi$ we use a 0-1 decision variables z_C which indicates whether the proposal is selected (1) or not (0). We include constraints that allow only one proposal to be selected from each agent. However, since the type restrictions are satisfied by all of the generated proposals, we do not need to include constraints to enforce this restriction. Although each proposal is generated so that each item is contained in at most one bid in that proposal, proposals from different agents may include the same items. Therefore we require a constraint that ensures that the winning combination of proposals includes each item at most once. We let $\overline{S}_C$ be the set of items in the bids in proposal C, that is, $\overline{S}_C = \cup_{k \in C} S_k$.

The set of winning bids can be determined by solving the following integer program:

$$Max \sum_{C \notin \Phi} w_C z_C.$$

Subject to

$$\sum_{C \in \Phi_p} z_C \leq 1 \qquad \text{for all} \quad p \in P \tag{6}$$

$$\sum_{C \in \Phi, i \in \overline{S_C}} z_C \leq 1 \qquad \text{for all} \quad i \in I \tag{7}$$

$$z_C \in \{0, 1\} \qquad \text{for all} \quad C \in \Phi. \tag{8}$$

Constraint (6) says that at most one proposal from each agent is in the winning combination. Constraint (7) says that each item is in at most one selected proposal. If each item must be sold, then the inequality in (7) should be changed to an equality. Potential infeasibility is addressed as in the previous formulation, by including a dummy agent and a dummy proposal for each individual item. Constraints (8) enforces integrality on the decision variables. Although this formulation may potentially have far more decision variables than the original formulation, our computational experience with this new formulation indicates that for moderate size problems it can be solved with minimal computational effort. This faster computation time may be due in part to the fact that all of the constraints have the same structure (sum of a set of variables is less than or equal to 1), and that the variables partition naturally in to sets of variables associated with each player. This partition, into so-called "special ordered sets," can be taken advantage of in commercial integer programming software.

Once a winning combination $Q \subset \Phi$ has been determined, we can extend this formulation to check to see whether another solution with equal revenue exists by adding a constraint to the model to prevent the proposals in Q from all being selected simultaneously.

$$\sum_{C \in Q} z_C \leq |Q| - 1. \tag{9}$$

In the case of duplicate proposals of differing types, we augment the set Q to form $Q' = Q \cup \{C \in \Phi \text{ s.t } \bar{S}_C = \bar{S}_{C'} \text{ and } w_C = w_{C'} \text{ for some } C' \in Q\}$ by adding proposals that have the same underlying sets and values, and replace Q by Q' in constraint (9). If the objective function value of this augmented integer program is the same as the objective function value of the original integer program, then we have a tie solution. Otherwise the winning solution is unique. Alternatively, we can use the "find all optimal solutions" function of OSL to find all optimal integer solutions to the original problem.

For any bid that is not in the winning solution, we can determine the surrogate value for the bid by solving an integer program in which the bid is forced in to the winning solution. Let b be the bid in question, and let p be the agent placing the bid. To formulate this integer program IP(b), we restrict Φ_p to proposals containing the bid b, and we change the special ordered set inequality for agent p to equality. We also eliminate all other proposals from all other agents that contain any of the licenses in the bid b. The surrogate value for the bid b is the optimal value of the original IP, minus the optimal value of IP(b). The resulting IPs have far fewer constraints than the original IP and solve relatively quickly; however there are a large number of them. The reason for the fast solution time is, at least in part, attributable to the fact that the number of non-integer elements in a solution is bounded by the number of constraints. The solution to one IP may provide a lower bound on the solution of other IPs, but beyond this obvious connection we have not found a way to speed the solution of this collection of problems.

Our limited computational experience to date is based on sample data obtained from the FCC to test a solver for their upcoming auction 31. The FCC provided seven data sets, four of which appeared designed to test specific features of the solver, and two of which appeared to represent realistic bidding. We used the largest data set to synthesize additional data, by increasing the number of agents and items, and by increasing the density of items per bid.

Problem id	# agents	# items	# types	Avg # bid /player/type	Bid value Deviation
Easy1	27	12	1	8.00	0.199
Easy2	28	12	2	9.27	0.455
Easy4	29	12	4	7.86	0.599
Easy10	29	12	10	4.55	0.626
Easy16	29	12	16	3.12	0.622
Base1	16	12	1	9.81	0.146
Base2	18	12	2	8.28	0.181
/6/0/0.5	24	12	2	7.17	0.188
/12/0/0.5	30	12	2	6.37	0.196
/18/0/1.0	36	12	2	5.64	0.199
/6/0/1.0	24	12	2	7.90	0.199

/12/0/1.0	30	12	2	7.97	0.189
/18/0/1.0	36	12	2	8.03	0.178
/18/16/1.0	36	16	2	8.03	0.483
/18/24/1.0	36	20	2	8.03	0.487

For these problems we generated all valid proposals. Surprisingly, the LP relaxation of the IP formulation yielded integer solutions in all instances. The column labeled LP solve time indicates the total CPU time to set up and solve the IP on a 500Mhz Intel-based laptop computer using Version 3.0 of OSL. For each losing bid, we also solved the LP obtained by forcing that bid into the solution. The number of these surrogate problems ranged from 7 to 266 for the different data sets. A very large percentage of the surrogate problems also yielded integer solutions to the LP relaxation, and those for which the LP solution was fractional typically required remarkably few branch and bound nodes to reach the optimal integer solution.

Problem id	# proposals	LP cpu sec	Surrogate value problems	Avg cpu Per surrogate	% non int
Easy1	36722	1.5	205	1.58	0.00
Easy2	49777	2.4	235	1.83	0.00
Easy4	57799	2.8	169	1.95	0.00
Easy10	60003	2.7	36	2.19	0.00
Easy16	60004	2.1	7	3.29	0.00
Base1	23845	1.0	152	1.04	18.42
Base2	29116	1.4	135	1.50	33.33
/6/0/0.5	19531	1.2	156	1.21	39.10
/12/0/0.5	13761	0.9	178	0.85	35.97
/18/0/1.0	13640	0.9	185	0.81	29.19
/6/0/1.0	30485	1.9	180	1.79	32.11
/12/0/1.0	33324	1.9	226	2.26	20.35
/18/0/1.0	38570	2.3	266	2.62	26.69
/18/16/1.0	38570	3.0	260	2.04	0.00
/18/24/1.0	38570	3.2	260	2.59	13.46

These small problems were surprisingly easy to solve. Additional examination of these and other data sets is required to understand whether the ease of solution is a result of the proposal–based formulation, or due to some other aspect of the problem.

3. Larger problems. For larger auctions, we propose a column generation approach in which only a subset of the feasible proposals are considered. The bids are used to generate an initial set of proposals. This initial set of proposals should, if possible include at least one proposal for each agent and at least one proposal for each item. In general, the initial set of proposals should include proposals that are of high value, relative to their contents. If information about the relative value of each item is available, this can be used to select combinations. Otherwise, one can make various estimates for the value of an item based on the value and number of items of each bid that contains the item. For each item $i \in I$ let n_i be the number of bids that contain i. If $n_i = 0$, then there are no bids that contain the item and clearly its value can be estimated to be 0. Otherwise one can estimate the value of i, denoted π_i, by averaging the value per item ratio of the bids that contain i:

$$\pi_i = \frac{1}{n_i} \sum_{k \in B, i \in S_k} \frac{v_k}{|S_k|}.$$

Using such an estimate for the value of each item, and defining the "excess" value of a bid to be the value of the bid minus the sum of the estimated value of the items in the bid, we compute "excess values" of a bid as $\tilde{v}_k = v_k - \sum_{i \in S_k} \pi_i$. Using these values, together with a greedy algorithm, one can quickly compute an initial set of proposals that have high "excess" value. Various methods can be used to ensure that there is at least one proposal for each item (that has a bid) and at least one proposal for each agent or agent-type combination. If each item must be sold, then for each item select the bid containing that item that has highest excess value, and add the proposal consisting of only that bid to the set Φ of proposals. If at least one proposal from each player must be considered, then for each player (or for each player, type combination) we generate the proposal consisting of the bid from that agent (or agent-type combination) with the highest excess value and add that proposal to Φ. To generate additional high value proposals, consider only the bids with positive excess cost. For each agent p and each type t, sort the positive excess value bids in $B_{p,t}$ in order of decreasing excess cost, $l(1), l(2), ..., l(n)$. Generate the proposal $C = \{l(1)\}$ and add it to the set of proposals. For each bid $k = 2, 3, ..., n$ such that $\bar{S}_C \cap S_{l(k)} = \phi$, set $C = C \cup \{l(k)\}$ and add the proposal C to the set Φ. If additional bids are required, a randomized greedy strategy, in which the bids are considered in other orders, or in which eligible bids are added to the proposal according to some probability distribution, can be used.

Once a sufficient number of proposals have been generated, the IP corresponding to these proposals are constructed. We solve only the linear programming relaxation of this IP, and retrieve the dual variables associated with each constraint. For an item i, we can use the dual variable π_i of the corresponding constraint (7) as an estimate of the value of the item. Using these new estimates, we can again compute the "excess" value of bids. We use the dual variables λ_p associated with the agent constraints as a threshold for the acceptance of a new proposal. A proposal C for agent p can increase the objective function value of the linear programming relaxation of the integer program if and only if

NISAN, N., "Bidding and Allocation in Combinatorial Auctions," presented at the 1999 Conference "Northwestern's Summer Workshop in Microeconomics," downloaded from cs.huji.ac.il/~noam/auctions.ps.

ROTHKOPF, M., A. PEKEC, AND R.M. HARSTAD, "Computationally Manageable Combinatorial Auctions," *Management Science,* 1998, **44,** 1131–1147.

TSUKIYAMA, ARIYOSHI AND IDE, SHIRAKAWA, in "A New Algorithm for Generating all the Maximal Independent Sets," *Siam Journal on Computing*, 1977, **6**, 505–517.

PRICE NEGOTIATIONS FOR PROCUREMENT OF DIRECT INPUTS

ANDREW J. DAVENPORT* AND JAYANT R. KALAGNANAM*

Abstract. Strategic sourcing relates to the procurement of direct inputs used in the manufacture of a firm's primary outputs. Such transactions are usually very large (in total quantity and the dollar value) and require the use of special price negotiation schemes that incorporate the appropriate business practices. In this paper, we present two auction mechanisms that have been developed for price negotiations in the context of strategic sourcing for a large food manufacturer. The first auction solicits supply curves as bids for procuring forecast demand for a direct input over a long planning horizon (such as a quarter). Typically suppliers provide *volume discounts* for such large orders and we call such auctions *volume discount auctions.* The second auction aggregates short-term (weekly) demand over direct inputs across multiple manufacturing plants in an effort to increase the total transaction size and allows suppliers to provide bundled all-or-nothing bids. Such auctions are called *combinatorial auctions.* A fundamental consideration that arises in the design of these schemes is the incorporation of business rules that appear as side constraints in the mathematical formulation for determining the winning bids (called the *winner determination problem*). We model the winner determination problem for these auctions as integer programs and provide some computational results using commercial integer programming solvers to illustrate the efficacy of these formulations.

1. Introduction. Strategic sourcing relates to the procurement of direct inputs used in the manufacture of a firms primary outputs. For example, the primary inputs for the manufacture of computers are processors, RAM (random access memory), hard drives, monitors, etc. Typically up to 90% of a firm expenditures are related to procuring direct inputs and as a result these transactions are large in volume as well as in dollar amount. As a result there is considerable room for price negotiations. However, a fundamental concern in such sourcing decisions is related to the reliability of suppliers, since defaulting suppliers might have considerable impact on the firm's ability to satisfy demand obligations. As a result these negotiations are generally confined to a restricted number of pre-certified suppliers having established relationships with the company.

The total quantity of direct inputs that needs to be procured is usually based on the forecasted demand for a planning horizon, typically a quarter. However, since there is considerable uncertainty in the forecast, a strategic decision is made regarding the fraction of the forecast demand that is procured up front[1]. The short-term (weekly) demand fluctuation over base level (which is procured using a long-term contract) is procured on a

---

*IBM T.J. Watson Research Center, P.O. Box 218, Yorktown Heights, New York 10598, USA. Email: davenport@us.ibm.com, jayant@us.ibm.com. IBM Technical Report 98774.

[1]This decision is made based on the uncertainty in the demand and price forecast and typically attempts to optimize some expected benefit. This decision problem is outside the scope of this paper.

weekly basis. However, the short-term demand is usually much smaller and there is less room for price negotiations. An approach adopted to address this is to aggregate demand over several commodities and over different locations and negotiate price for the entire bundle. In addition, in order to exploit the cost complementarities that suppliers might have for different commodities or locations it becomes necessary to allow all-or-nothing bids over bundles.

In this paper auction mechanisms are proposed for price negotiations related to the procurement of both the long-term and short-term demand. A *volume discount* auction is proposed for the procurement of large volume (long-term) demand. Volume discount auctions are a generalization of the multi-unit auctions [Ausubel, 1997] in that they allow bidders to specify different price levels for different quantities[2]. Typically for large volume price negotiations, suppliers provide volume discounts, i.e. as the total quantity bought increases the price drops monotonically. *Combinatorial auctions* are proposed for the procurement of short-term demand. Combinatorial auctions allow the buyer to procure a variety of commodities simultaneously, and allow the suppliers to bid on sets or bundles of items. Because of the complementarities (or substitution effects) between the production or transportation costs of different commodities, suppliers can provide lower prices on sets of commodities.

The focus of this paper is the *winner determination problem*: given a set of bids for some specified demand, which subset of the bids should the buyer accept so as to minimize the total procurement cost for this total demand. A fundamental consideration in formulating these price negotiation schemes is the business rules that firms use to constrain their selection of suppliers. Typical rules are motivated by risk hedging considerations. For example the number of winning suppliers chosen is constrained to have a minimum since dependence on too few suppliers might expose the firm to the misfortunes or supply fluctuations of the chosen suppliers. On the other hand, too many suppliers would lead to high administrative costs and hence the number of winning suppliers is also constrained from above. Other constraints emerge from limits that firms impose on the total volume allocated to transactions with each supplier. We provide mathematical formulations for both the volume discount auction and the combinatorial auction taking into account the business rules as side constraints. The side constraints fundamentally alter the structure of the problems for both cases and lead to novel formulations that have not been studied in the auctions literature [de Vries & Vohra, 2000].

The price negotiation mechanisms discussed in this paper have been developed and deployed at a large food manufacturer for their strategic sourcing operations. The auctions are implemented in the context of a pri-

[2]Notice that auctions in the context of procurement are reverse auctions that are buyer driven.

vate business-to-business electronic exchange (B2B exchange). Such private electronic markets increasingly seem to be the emergent model for most large firms.

Since the focus of this paper is on the decision problems that lie at the core of the price negotiation scheme we first provide a general description of the volume discount auction and the combinatorial auction problem and the optimization models that need to be solved to determine the winners for a given set of bids. Section 2 provides a description of the volume discount auction, the business rules and the associated mathematical formulation of the winner determination problem as a linear integer program. Section 3 provides a description of the reverse combinatorial auction, the business rules and the mathematical formulation of the winner determination problem. Computational results investigating the impact of the side constraints on solving the winner determination problem for randomly generated data sets are also presented in these two sections. Section 4 discusses some modeling issues that are required to capture some of the operational details of the auctions. Section 5 provides a discussion of some of the research issues that arise from this work.

2. Volume discount auctions. Volume discount auctions are used in a procurement context where there is a single buyer and multiple sellers. A buyer wishes to purchase some quantity of an item (which we also refer to as *lots*), and invites sellers to submit bids for these items. Bids in a volume discount auction allow the seller to specify the price they charge for an item as a function of quantity that is being purchased. For instance, a computer manufacturer may charge $1000 per computer for up to 100 computers, but for each additional computer over 100 computers would charge only $750 per computer. Bids take the form of *supply curves*, specifying the price that is to be charged per unit of item when the quantity of items being purchased lies within a particular quantity interval.

2.1. Example. We consider an example of a volume discount procurement auction where the buyer wishes to purchase 60 units of some commodity. Figures 1(a) and (b) illustrate two supply curves that could be provided by two different sellers bidding in this auction. Each supply curve consists of a list of pairs, composed of a price and a quantity interval. Each pair specifies the price that is being offered per unit of commodity, if the buyer purchases a number of units within this quantity interval. For instance, in supply curve 1 (Figure 1(a)) the buyer is offering a price of $100 per unit, for quantities between 1 and 20 units. If the buyer is willing to buy more units from this seller, say 25 units, then they can get a better price per unit, in this case $45 per unit for each additional unit over 20.

The winner determination problem for the volume discount auction is to select a set of winning bids, where for each bid we select a price and quantity of items to be bought, such that the total demand of the buyer is satisified at the lowest price.

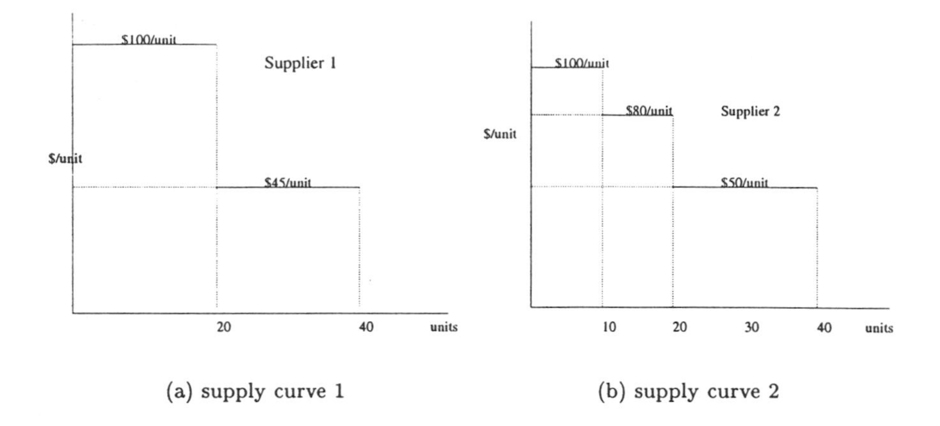

(a) supply curve 1 (b) supply curve 2

FIG. 1. *Example supply curves.*

The winner determination problem for this type of auction is straightforward if there are no constraints on the maximum number of units that can be purchased from a single seller, i.e. the total quantity of the commodity requested by the buyer can be met by a single seller. In this case, it suffices to determine the best price that each seller is offering for the quantity requested by the buyer, and buy the entire quantity from this seller. For instance, if the buyer has a demand for 30 units, and we are given the supply curve bids as shown in Figure 1, then we can determine the best price is given by seller 2 at a total cost of $2300 per unit for the entire demand.

For the particular project we were involved with, there were constraints on both the minimum and maximum number of units that the buyer may purchase from each seller. Such a constraint may state that the maximum quantity that can be purchased from a single seller is 40 units. If the buyer wishes to buy more than 40 units, the winner determination problem becomes more complex. For instance in our example, if the total demand is 60 units, the best solution is to buy 30 units from seller 2 for a cost of $2300, and 30 units from seller 1 for a cost of $2450, resulting in a total cost of $4750. By buying 40 units from seller 1 we could get more units at a lower price of $50 per unit (cost $2800), however this would decrease the quantity, and thus increase the price per unit of the items we buy from seller 2 for a cost of $2000 resulting in a total cost of $4800.

2.2. Mathematical formulation. We represent a volume discount procurement auction in the following way:

1. The buyer has K lots that s/he needs to procure, and requires a quantity $Q^k, k = 1, \ldots, K$ for each lot;

2. The buyer also identifies a list of potential suppliers $i = 1, \ldots, N$ that can bid in the auction;
3. Each supplier responds with a bid composed of a supply curve (at most one for each lot). A supply curve from supplier i for lot k given by a bid B_i^k consists of a list of M_i^k price-quantity pairs, $\{(P_{i1}^k, [Q_{i1,low}^k, Q_{i1,high}^k]), \ldots, (P_{i|M_i^k|}^k, [Q_{i|M_i^k|,low}^k, Q_{i|M_i^k|,high}^k])\}$. Each price-quantity pair $(P_{ij}^k, [Q_{ij,low}^k, Q_{ij,high}^k])$ specifies the price P_{ij}^k that the supplier i is willing to charge per unit of the lot k if the number of units bought from this supplier lies within the interval $[Q_{ij,low}^k, Q_{ij,high}^k]$. We assume that the quantity intervals within a single supply curve are all pairwise disjoint.

The winner determination problem for the volume discount auction can be formulated as a mixed integer programming problem (MIP) in the following way:

1. We associate a decision variable x_{ij}^k to each price-quantity pair $(P_{ij}^k, [Q_{ij,low}^k, Q_{ij,high}^k])$ for each bid B_i^k. This variable takes the value 1 if we buy some number of units of the lot through this bid within the quantity range $[Q_{ij,low}^k, Q_{ij,high}^k]$ at the price P_{ij}^k; it takes the value 0 otherwise.
2. We associate a continuous variable z_{ij}^k with each price-quantity pair, which specifies the exact number of units of the lot that is to be purchased from the bid B_i^k within this price-quantity pair.

The MIP formulation is then given as follows:

$$\text{minimize} \quad \sum_{k \in K} \sum_{i \in N} \sum_{j \in M_i^k} (z_{ij}^k\, P_{ij}^k + x_{ij}^k\, C_{ij}^k)$$

$$\text{subject to}$$

$$(1) \qquad \begin{aligned}
& z_{ij}^k - (Q_{ij,high}^k - Q_{ij,low}^k)\, x_{ij}^k \leq 0 && \forall i \in N, \forall j \in M_i^k && (a) \\
& \sum_{j \in M_i^k} x_{ij}^k \leq 1 && \forall i \in N, \forall k \in K && (b) \\
& \sum_{i \in N} \sum_{j \in M_i^k} (z_{ij}^k + x_{ij}^k\, Q_{ij,low}^k) \geq Q^k && \forall k \in K && (c) \\
& x_{ij}^k \in \{0, 1\} && \forall i \in N, \forall j \in M_i^k, \forall k \in K \\
& z_{ij}^k \geq 0 && \forall i \in N, \forall j \in M_i^k, \forall k \in K.
\end{aligned}$$

The coefficient C_{ij}^k is a constant and computed apriori as:

$$(2) \qquad C_{ij}^k = \sum_{\hat{j}=1}^{j-1} P_{i\hat{j}}^k (Q_{i\hat{j},high}^k - Q_{i\hat{j},low}^k).$$

Constraints (a) specifics for each price-quantity pair $(P_{ij}^k, [Q_{ij,low}^k, Q_{ij,high}^k])$, that if we buy some quantity of the commodity from the bid at the price P^k_{ij}, this quantity must lie within the range $[Q^k_{ij,low}, Q^k_{ij,high}]$. Constraint (b) specifies that for each winning bid, we can only buy at a price and quantity that corresponds to a single price-quantity pair. Constraint (c) states that we must determine a winning set of bids such that the total demand of the buyer for each lot k is satisfied.

We have included all lots that the buyer wishes to procure in this formulation, even though, as it stands, we could solve the winner determination problem for each lot independently. This is done for convenience: in the next section we introduce side constraints which involve all the lots in the auction.

2.3. Side constraints. In a real world setting there are several considerations beside cost minimization. These considerations often arise from business practice and/or operational considerations and are specified as a set of constraints that need to be specified while picking a set of winning suppliers. We discuss four such business rules/constraints that we encountered in our application with the food industry.

2.3.1. Number of winning suppliers. An important consideration while choosing winning bids is to make sure that the entire supply is not sourced from too few suppliers, since this creates a high exposure if some of them are not able to deliver on their promise. On the other hand, having too many suppliers creates a high overhead cost in terms of managing a large number of supplier relationships. These considerations introduce constraints on the minimum, S_{min}, and maximum, S_{max}, number of winning suppliers in the solution to the winner determination problem.

These constraints are encoded in the MIP formulation in the following way. We introduce an indicator variable y_i for each supplier i, which takes the value 1 if the supplier has any winning bids and 0 otherwise. The first constraint sets y_i to 1 if supplier i has any winning bids. Note that the constant multiplier ensures that the right hand side is large enough when y_i is one[3].

$$
\begin{aligned}
&\sum_{k\in K}\sum_{j\in M_i} x^k_{ij} \leq y_i\, K && \forall i \in N\\
&y_i \in \{0,1\} && \forall i \in N \qquad (3)\\
&S_{min} \leq \sum_{i\in N} y_i \leq S_{max}.
\end{aligned}
$$

2.3.2. Local lot level constraints. Local lot level constraints state for each supplier i and lot k, the minimum $q^k_{i,min}$ and maximum $q^k_{i,max}$ quantity that can be allocated to supplier of this lot type. For instance, a constraint may state that some supplier must be allocated at least 500 tons and at most 12,000 tons of a particular lot. These constraints are once

[3]The constant multiplier K follows from constraint (c) in the formulation.

again motivated by concerns similar to the ones related to the number of winning suppliers. Rather than add these constraints to the formulation, they can be used to prune the supply curves of the bids from each supplier so that they lie within a feasible range with respect to this constraint.

2.3.3. Global lot level constraints. For each supplier i, the buyer provides bounds on the total allocation, across all lots, to lie between $W_{i,min}$ and $W_{i,max}$. In the winning allocation, if the supplier i has *any allocation at all* then it must lie in this range. These constraints can be expressed in the following way in the formulation:

$$
(4) \qquad \begin{aligned}
& y_i W_{i,min} - \sum_{k \in K} \sum_{j \in M_i^k} (z_{ij}^k + x_{ij}^k Q_{ij,low}^k) \leq 0 \quad \forall i \in N \\
& \sum_{k \in K} \sum_{j \in M_i^k} (z_{ij}^k + x_{ij}^k Q_{ij,low}^k) - y_i W_{i,max} \leq 0 \quad \forall i \in N.
\end{aligned}
$$

Note that these constraints imply the first constraint in Equation 3.

2.3.4. Reservation prices. A maximum allowable price per unit can be provided by the buyer for each lot. This constraint is not encoded in the MIP formulation for the winner determination problem. Instead bids, or portions of the supply curve in each bid, that fail to satisfy this constraint are disregarded by the solver engine.

2.4. Relevant literature. The restriction of the volume discount auction winner determination problem to a single lot can be modeled as a variation of the *multiple choice knapsack* problem [Martello & Toth, 1989], and as such is NP-hard.

2.5. Auction mechanism and computational issues. The buyer in the auction specifies:

1. the lot to be purchased, required quantity and a reservation price;
2. a list of acceptable suppliers for each lot;
3. lot level constraints (discussed below);

Each supplier submits a supply curve bid for each lot. The supplier may submit up to 10 price-quantity pairs within each bid, ranging from their minimum to maximum supply limits. The supplier may choose to submit bids over a smaller quantity range than is required by the buyer (e.g., the minimum required volume is 200 tons, but the supplier chooses only to place bids for quantities greater than 300 tons.)

This is a multi-round sealed bid auction. Once the bids have been submitted, the system processes the bids, solves the winner determination problem and returns a minimum cost optimal solution to the buyer. Suppliers are informed of the winning bids in each round, and may submit new bids. Typically there are up to 30 lots and about 10 suppliers for each lot, and the optimization engine needs to find a solution within a couple of minutes.

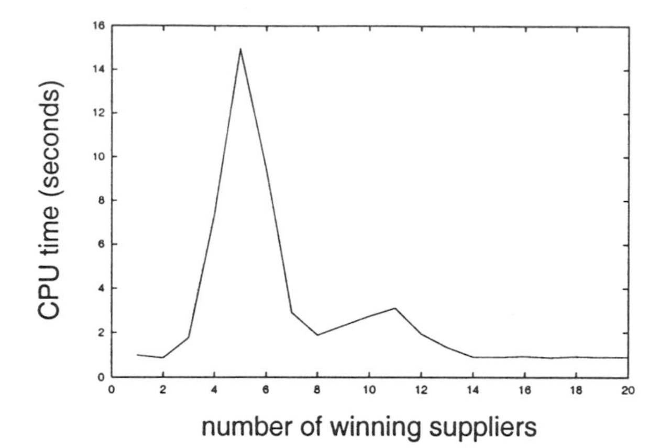

FIG. 2. *Experimental results for a volume discount auction, varying the number of allowed winning suppliers constraint.*

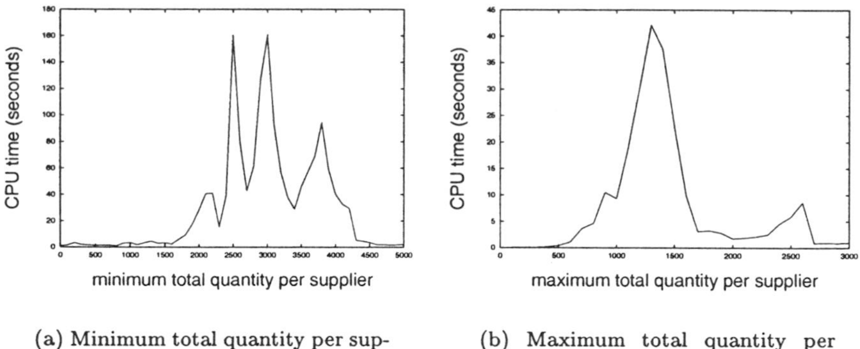

(a) Minimum total quantity per supplier constraint

(b) Maximum total quantity per supplier constraint

FIG. 3. *Experimental results for a volume discount auction, varying the constraints on total quantity that can be awarded to each supplier.*

A dynamic programming approach can be used to solve this problem. We found however that commercial integer programming software using a branch-and-bound approach was able to solve problems specified by the customer on the order of seconds or minutes.

Figures 2 and 3 present CPU time results (in seconds) for solving a randomly generated instance of the volume discount winner determination problem, where we vary side constraints. This problem had 10 lots available in the auction, with 100 suppliers submitting up to 5 bids each. Figure 2 illustrates how varying the number of winning suppliers affects problem difficulty. We set the minimum number of winning suppliers to be equal to the maximum number of winning suppliers, which we varied between 1 and 20. No other side constraints were set for this experiment. The optimal allocation for this problem, without any side constraints on the number of winning suppliers, has 13 winning suppliers. The problem becomes harder to solve as we decrease the number of permitted winning suppliers below this number. For very small numbers of winning suppliers, the problem becomes easier to solve, since with few suppliers it may not be possible to find an allocation which satisfies the total demand of the buyer. Note that the results presented here were generated using an instance that uses a slightly different objective function than the one presented in Equation 1. However, the qualitative behavior presented here persists across both models.

Figure 3 investigates the minimum and maximum total quantity assigned to each supplier constraint. Once again, no other side constraints were set for this experiment. Here the optimal allocation, without any side constraints, gives a minimum quantity of 100 and a maximum quantity of 2700 to some winning suppliers[4]. We see here that deviating from this optimal allocation, by decreasing the maximum or increasing the minimum quantity that can be assigned to each supplier, makes the problem harder to solve.

3. Combinatorial auctions. Combinatorial auctions provide a mechanism whereby bidders can submit bids on combinations of items [Rassenti et al., 1982, Fujishima et al., 1999, Sandholm, 1999, de Vries & Vohra, 2000]. Such auctions are used when, due to complementarities or substitution effects, bidders may have preferences not just for particular items but also for sets, or bundles, of items. In this paper, we describe a reverse combinatorial auction. The auction is run as a procurement auction, where the buyer wishes to purchase different items of varying quantities, for the cheapest overall price. The total quantity of each item is called a *lot* and is treated as indivisible unit of some weight. Suppliers can bid on combinations of items, however, a bid on any item has to be for the entire lot for that item. The winner determination problem for the single unit reverse combinatorial auction problem is to select a winning set of bids such that each item is included in at least one winning bid, and the total cost of procurement is minimized. This problem is a weighted set covering problem, which is known to be NP-hard.

3.1. Mathematical formulation. We are given a set of K items, where for each item $k \in K$ there is a demand for d^k units of the item (called a lot). Each supplier $i \in N$ is allowed up to M bids indexed by j. We associate with each bid B_{ij} a zero-one vector a^k_{ij}, $k = 1, \ldots, K$ where

[4]Note that the allocation found for both the combinatorial and volume discount auction winner determination problems may not satisfy the total demand of the auction. This may not be possible with the side constraints, in which case the mechanism described in Section 3.5 is used to return a partial allocation.

$a_{ij}^k = 1$ if B_{ij} will supply the (entire) lot corresponding to item k, and zero otherwise. Each bid B_{ij} offers a price p_{ij} at which the bidder is willing to supply the combination of items in the bid. A mixed integer programming formulation for the reverse combinatorial auction can be written as follows:

$$
(4) \quad \begin{array}{lll}
\textit{minimize} & \displaystyle\sum_{i \in N} \sum_{j \in M} p_{ij} x_{ij} & \\
\textit{subject to} & & \\
& \sum_{i \in N} \sum_{j \in M} a_{ij}^k x_{ij} \geq 1 \quad \forall k \in K & (a) \\
& x_{ij} \in \{0,1\} \qquad\qquad\qquad \forall i \in N, \forall j \in M. &
\end{array}
$$

The decision variable x_{ij} takes the value 1 if the bid B_{ij} is a winning bid in the auction, and 0 otherwise. Constraint (a) states that the total number of units of each item in all the winning bids must satisfy the demand the buyer has for this item. In this auction goods complement each other, so the valuations of sets of items are sub-additive. That is, the price offered by a particular supplier for two non-disjoint sets of items A and B, $p(A)$ and $p(B)$, is such that $p(A) + p(B) \geq p(A + B)$. Such sub-additive cost functions result from complementary costs such as the use of a common warehouse, or unused capacity in a carrier.

3.2. Side constraints. The auction mechanism and side constraints are similar to those of the volume discount auction. The side constraints are:

1. minimum and maximum number of winning suppliers;
2. minimum and maximum total quantity allocated to each supplier;
3. reservation prices on each lot.

These constraints can be added to the MIP formulation as follows:

$$
(5) \quad \begin{array}{lll}
W_{i,min}\, y_i \leq \displaystyle\sum_{k \in K} \sum_{j \in M} a_{ij}^k d^k x_{ij} & \forall i \in N & (a) \\
\displaystyle\sum_{k \in K} \sum_{j \in M} a_{ij}^k d^k x_{ij} \leq W_{i,max}\, y_i & \forall i \in N & (b) \\
\displaystyle\sum_{j \in M} x_{ij} \geq y_i & \forall i \in N & (c) \\
S_{min} \leq \displaystyle\sum_{i \in N} y_i \leq S_{max} & & (d) \\
y_i \in \{0,1\} & \forall i \in N. &
\end{array}
$$

$W_{i,min}$ and $W_{i,max}$ relate to the minimum and maximum quantity that can be allocated to any supplier i. Constraints (a) and (b) restrict the total allocation to any supplier to lie within ($W_{i,min}$, $W_{i,max}$). Note that y_i is an indicator variable that takes the value 1 if supplier i is allocated any lot. Notice that if $W_{i,min} = 0$ then y_i becomes a free variable. In order to fix this we introduce constraint (c) which ensures that $y_i = 0$ if no bids from supplier i are chosen. S_{min} and S_{max} relate to the minimum and maximum number of winners required for the allocation. Constraint (d) restricts the total number of winners to be within the range (S_{min}, S_{max}).

A typical problem is specified to have 6–10 suppliers and 250 lots for auction. The maximum problem size is given as 30 suppliers and 400 lots. There is no limit on the number of lots in each bid, although this could be limited if required to improve solution time. Solutions for the winner determination problem are needed on the order of 5–10 minutes.

3.3. Relevant literature. The formulation (4) provided for the reverse combinatorial procurement auction is a set covering problem, which is NP-hard [Garey & Johnson, 1979]. Since this remains the core even after adding the side constraints, the overall formulation with side constraints remains NP-hard. There is a greedy approach for set covering which yields an $O(\log_2 n)$ approximation, and it has been shown that this cannot be significantly improved [Shmoys, 1995].

The addition of the side constraints makes a fundamental impact on the feasibility of the problem. Without these we could always choose all bids and get a feasible (albeit) expensive solution. However, with limits on the quantity allocated to each supplier and on the total number of winners, a feasible solution might not exist or if one exists it might be difficult to find.

3.4. Auction mechanism and computations. We ran some experiments to investigate the difficulty of solving the winner determination problem with respect to the side constraints in the auction. Figure 4 presents results for one example of a single round reverse combinatorial auction winner determination problem. This particular instance was generated randomly. For each supplier we generated a set of lots the supplier was interested in, and a set of bids for different subsets of this set. A single bundled bid for a set of lots $\mathcal{S}$ would be for a lower price than that of the sum of the prices of any set of bids by the same supplier which also, in total, covers all the lots in $\mathcal{S}$[5]. This particular problem had 30 suppliers and 15 lots available for auction. Each supplier made multiple bids: in total there were 352 bids. With no side constraints, there were 12 winning bids and 12 winning suppliers. Figure 4(a) illustrates how varying the number of permitted winning suppliers affects problem solving time. Once again, we set the minimum number of winning suppliers to be equal to the maximum number of winning suppliers, which was varied between 1 and 30. Decreasing the number of permitted winning suppliers below 14 increased problem solving time significantly, reaching a peak at 4 winning suppliers. The total demand of the auction could not be satisfied with less than 5

[5]This property, which should hold for bids in a procurement auction, would not hold for the combinatorial auction problems discussed in [Sandholm, 1999, de Vries & Vohra, 2000].

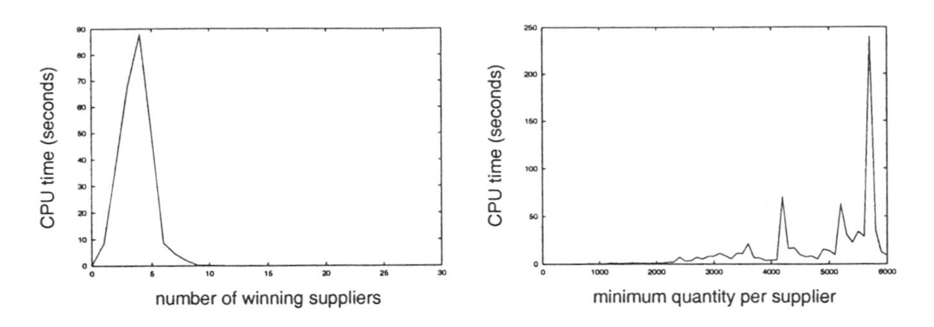

(a) Varying the permitted number of winning suppliers constraint (minimum = maximum)

(b) Varying the total minimum quantity per supplier constraint

FIG. 4. *Experimental results for a combinatorial auction with varying side constraints.*

suppliers. Increasing the number of permitted suppliers did not make the problem harder to solve. Figure 4(b) shows what happens when we vary the minimum quantity assigned to each supplier constraint. For these experiments, each supplier has the same minimum total quantity constraint. With no side constraints, the solution to the winner determination problem assigns a maximum quantity of 1800 to any supplier. We see here that as we increase this minimum quantity, the problem becomes increasingly harder to solve. Conversely, we found that problem difficulty was not very much affected by varying the maximum quantity allocated to each supplier constraint.

3.5. Feasibility. As discussed earlier, finding feasible solutions for some instances of this problem is difficult. This also has a big impact on the run time. A common trick for dealing with this situation is to introduce a dummy bid for each lot with a price set to K times the largest bid price in the auction. Dummy bids are added to the integer programming formulation for the demand cover constraints, but do not appear in the side constraints for min/max quantity or min/max number of winning suppliers. A dummy bid for a particular lot type will only appear in an optimal solution to the winner determination problem for these auctions if there is no way to satisfy the demand for the lot using an existing real bid. Thus any solution to the winner determination problem will still satisfy the entire demand for lots in the auction. Should there be dummy bids in the set of winning bids for the winner determination problem, these are removed from the set before the real winning bids are returned to the buyer and infeasibility is indicated.

4. Operational issues. In this section we discuss two operational issues that emerged in this application. Both of these issues arise as requirements in a practical setting and have interesting implications in terms of refining the formulations provided above.

4.1. Feasibility. In a multi-round auction mechanism, there may not be enough bids placed in earlier rounds in the auction to satisfy the total demand requested by the procurer. In such cases, the formulations we have presented earlier for the volume discount and combinatorial auction winner determination problem (formulations (1) and (4)) will be infeasible. In practice, we found that the procurer preferred to receive a partial allocation if the entire demand could not be met[6]. We can model this requirement using the existing formulations by adding *dummy bids* to the formulation as described earlier.

4.2. Timestamps. One issue which arises in the context of multi-round auctions is the treatment of bids made in different rounds of the auction, for the same bundle of items at the same price or supply curve. Consider the following example: A combinatorial procurement auction is created to purchase some quantities of items $\{1,2,3\}$. In the first round of the auction Supplier 1 makes a bid B_1 for items $\{1,2,3\}$ at a price of \$100, and Supplier 2 bids B_2 of \$30 for item $\{1\}$. The solution to the winner determination problem for this round is that Supplier 1 wins with bid B_1. During the second round a new supplier, Supplier 3, enters the auction with a bid B_3 for items $\{2,3\}$ at \$70. Using the integer prgramming formulation presented earlier (Equation 4), there are two potential solutions to this combinatorial auction winner determination problem: either $\{B_1\}$ or $\{B_2, B_3\}$. In both cases the total cost to the procurer is \$100.

From a business point of view these two solutions are not both equally desirable. In a multi-round procurement auction, new bids should only supercede existing winning bids if they result in a lower price being paid for the items in the auction. The usual way to handle such situations is to have a rule stating that, given two identical bids, the bid that was made earlier in time is to be preferred. This rule is straightforward to enforce in a simple, single or multi-item forward or reverse auction. In the context of combinatorial and volume discount auctions this rule becomes harder to enforce, since the number of possible solutions to the winner determination problem may be exponential in the number of bids placed in the auction.

In practice, each bid B_i is associated with a timestamp TS_i, representing the time it was accepted into the auction[7]. The new objective for the winner determination problem then becomes to select the set from the set

---

[6]The customer specified that the side constraints be satisifed at all times, and only the demand constraints could be relaxed.

[7]For the purposes of this paper, we will regard the timestamp as being an integer. In the customer implementation of these auctions, the timestamp was of the SQL type 'timestamp'.

of all sets of bids which minimizes the total cost, such that this set minimizes the sum of the timestamps of all the winning bids[8]. That is, we have a further timestamp objective, which is secondary to the cost objective, which is:

$$\text{minimize} \sum_{i \in N} TS_i x_{ij}. \tag{6}$$

We have considered and implemented two ways of dealing with this issue. We describe these techniques below.

4.2.1. Multiple formulations. The first proposal to deal with the timestamp issue works by solving two integer programming problems. The first problem we solve is the winner determination problem (for either the combinatorial or volume discount auction) without considering timestamps at all, i.e., by using the formulations (1) and (4).

We then formulate a new integer programming problem, which differs from the first formulation in two ways:

1. We take the value of the objective v in the solution to the first integer programming problem, which represents the minimum cost price to the procurer to purchase their demand in the auction. We add as a constraint that the demand should be bought at this price. For the combinatorial auction formulation (4), this constraint is written as:

$$\sum_{i \in N} p_i x_{ij} = v. \tag{7}$$

For the volume discount auction formulation 1 this constraint is:

$$\sum_{i \in N} \sum_{j \in M_i^k} z_{ij}^k P_{ij}^k = v. \tag{8}$$

2. The new objective becomes to minimize the sum of the timestamps of winning bids (Equation 6).

4.2.2. Price modification. The second technique encodes information concerning the time a bid was made into the price of the bid that is seen by the winner determination solver. Typically, prices are expressed to a fixed number of decimal places, usually two. Thus within the representation of the bid price as a floating point number, we can use digits beyond two decimal places to encode timestamp information. For instance, in our previous combinatorial auction example we had three bids: B_1 of \$100 for items $\{1,2,3\}$, B_2 of \$30 for items $\{1\}$ and B_3 of \$70 for items $\{2,3\}$. If

[8]In practice we found it to be more efficient and scalable to sort the bids by their timestamps into a list, and use the order that bids appear in this list, rather than their timestamps, in the new objective.

these bids B_1, B_2 and B_3 were made at times 1, 2 and 4 respectively, we might encode the timestamps into the bid prices such at p_1 = \$100.001, p_2 =\$30.002 and p_3 = \$70.004. We then solve the integer programming problem with the same objective to minimize the total cost to the procurer. The set of winning bids that minimizes the total cost will also be the one whose bids were made the earliest. In this example, bid B_1 has a lower cost (100.001) than bids B_2 and B_3 combined (100.006), so B_1 will be the winning bid.

Some care must be taken when encoding timestamp information into bid prices. Firstly, using the scheme outlined above, the following situation may occur. Suppose that bid B_3 was accepted at time 2, B_2 accepted at time 4 and bid B_1 accepted at time 5. The corresponding prices on these bids with timestamps become p_3 = \$70.002, p_2 = \$30.004 and p_1 = \$100.005. Bid B_1 would be the winning bid in this example, even though bids B_2 and B_3 were made earlier and for the same total (real) cost. The problem here is that the sum of the timestamp portions of the bid prices for bids B_2 and B_3 is greater than that for bid B_1, even though their individual timestamp parts of the bid price are less. To resolve this problem we must take into account the number of items within each bid when determining timestamp bid price modifications. To do this, we sort all the bids by their timestamps into a list. We set a counter variable *ts = 0*. We then take each bid B_i from the list in order, increment the variable *ts* by the number of items in bid B_i, and set the timestamp portion of the bid B_i to the current value of the variable *ts*. For our running example, the bid prices become p_3 = \$70.002, p_2 = \$30.003 and p_1 = \$100.006.

The advantage of this price modification technique for dealing with timestamps is that we only have to solve one integer programming problem in order to determine the winners of the auction. The main disadvantage is the precision of the floating point arithmetic limits how many bids we can deal with in this way, while guaranteeing that the timestamp part of the modified bid price does not interfere with the real part of the modified bid price[9].

5. Discussion. The sell side forward combinatorial auction has received much attention in recent years [Rassenti et al., 1982, Fujishima et al., 1999, Sandholm, 1999, de Vries & Vohra, 2000], especially with respect to its economic properties and the computational difficulty of solving its winner determination problem. However, little has been said about the reverse combinatorial auction. We have presented details of an application of the reverse combinatorial auction to procurement of directs inputs, where short term demand is aggregated across multiple manufacturing plants, allowing suppliers to provide bundled all-or-nothing bids. In

[9]In practice, given n bids each for m items, we need ceil($\log_2(\frac{nm\times(nm+1)}{2})$) exclusive digits within the bid price in which to store the timestamp information, in order to guarantee that the timestamp part of the bid price does not interfere with the real (price) part of the bid price.

such a context, much of the work in combinatorial auctions may not be applicable to real world situations such as direct procurement, since they ignore side constraints arising as a result of business rules on how demand can be allocated across different suppliers. Such constraints not only impact the computational difficulty of finding an optimal solution to such problems (as illustrated in our empirical studies), but also make determining whether a feasible solution exists an NP-complete problem. In contrast, finding a feasible solution for the forward and reverse combinatorial auctions without side constraints is trivial.

The volume discount auction is common practice in industry for large transactions that are made over a long term time horizon. They provide a way for suppliers to give an indicator of their production costs without revealing details of their manufacturing and capacity constraints. Once again, this type of auction has received little attention in the research literature, and little is known about its computational and economic properties.

There are three functions a decision support system can provide in an auction setting: winner determination, price signalling and bid reformulation. This paper has concentrated on winner determination. In a multi-round auction, price signalling and bid reformulation refer to what information can be given to a losing bidder to enable them to make a winning bid in the next round of an auction. This is a significant research issue. The mechanism used for the project described in this paper informed all bidders of the winning bids in each round of the auction. This mechanism is weak, in the sense that it does not tell bidders how they should reformulate their bids for the next round. For the combinatorial and volume discount auctions, designing price signalling mechanisms to give bidders such information is complex, given the combinatorial nature of the problem.

Recent work [Bickchandani & Ostroy, 1998] has provided an extended formulation for the set packing problem which is integral. The dual of this formulation provides nonlinear prices (with price discrimination) for various bundles. Based on these properties it has been shown that efficient and incentive compatible iterative auctions can be designed [Parkes & Ungar, 2000, Bickchandani & Ostroy, 1998]. However, little attention has been paid to the set covering formulation with regard to designing efficient and incentive compatible mechanisms. In practice, the most likely use of combinatorial auctions seems to be in the context of procurement where the set covering formulation is central. The volume discount auction provides another interesting problem where identifying extended formulations that provide dual prices would be very useful. An additional direction for investigation is the impact of side constraints (the business rules used in practice) on the extended formulation even for the set packing case. The robustness of an auction design would greatly depend on whether the economic properties can be shown to hold under various business rules that appear as side constraints in practice.

Acknowledgements. We wish to gratefully acknowledge useful discussions with Brenda Dietrich, Marta Eso, David Jensen, and Laci Ladanyi on issues related to this paper.

REFERENCES

[Ausubel, 1997] AUSUBEL, L. (1997). An efficient ascending-bid auction for multiple objects Tech. rep., Dept. of Economics, University of Maryland.

[Bickchandani & Ostroy, 1998] BICKCHANDANI, S. & OSTROY, J.M. (1998). The package assignment model. Tech. rep., UCLA.

[de Vries & Vohra, 2000] DE VRIES, S. & VOHRA, R. (2000). Combinatorial auctions: a survey. Tech. rep., Dept of Managerial Economics and Decision Sciences, Kellog School of Management, Northwestern University, Evanston, IL 60208.

[Fujishima et al., 1999] FUJISHIMA, Y., LEYTON-BROWN, K., & SHOHAM, Y. (1999). Taming the computationally complexity of combinatorial auctions: optimal and approximate approaches. In *Proceedings of IJCAI'99*. Morgan Kaufmann.

[Garey & Johnson, 1979] GAREY, M.R. & JOHNSON, D.S. (1979). *Computers and Intractability: A Guide to the Theory of NP-Completeness.* W.H. Freeman and Company, New York.

[Martello & Toth, 1989] MARTELLO, S. & TOTH, P. (1989). *Knapsack problems.* John Wiley and Sons Ltd., New York.

[Parkes & Ungar, 2000] PARKES, D. & UNGAR, L.H. (2000). Iterative Combinatorial Auctions: Theory and Practice. In *Proc. 17th National Conference in Artificial Intelligence*, pp. 74–81.

[Rassenti et al., 1982] RASSENTI, S.J., SMITH, V.L., & BULFIN, R.L. (1982). A combinatorial auction mechanism for airport time slot allocation. *Bell Journal of Economics*, **13**(2), 402–417.

[Sandholm, 1999] SANDHOLM, T. (1999). An algorithm for optimal winner determination in combinatorial auctions. In *Proceedings of IJCAI'99*, pp. 542–547. Morgan Kaufmann.

[Shmoys, 1995] SHMOYS, D. (1995). Computing near-optimal solutions to combinatorial optimization solutions. In *Advances in Combinatorial Optimization*. AMS.

AN ITERATIVE ONLINE AUCTION FOR AIRLINE SEATS

MARTA ESO*

Abstract. Due to their low cost and ease of access, online auctions are a very popular way of selling perishable excess inventory in the travel industry. We analyze online auctions for airline seats where leftover seat capacity on flights between two given cities is traded. In addition to the number of tickets requested and the bid amount, bids may specify a *set* of alternative flights, each equally acceptable for the bidder. A winning bid will have all requested seats allocated on the same flights. We study an iterative mechanism where the decision to accept and reject bids and to provide minimum bid suggestion for rejected and displaced bids has to be made in real time. Each iteration of the mechanism can be thought of as a general combinatorial auction where customers bid on bundles of flights. We discuss heuristics as well as exact solution methods for solving the underlying Integer Program. We show that the model can be easily extended to incorporate more general settings. Preliminary computational results for synthetic data are also presented.

Key words. Mechanism design, Combinatorial auctions, Integer Programming.

1. Introduction. Due to their low cost and ease of access, online auctions are a very popular way of selling perishable excess inventory in the travel industry. Auctions for specific tickets, trips or even vacation packages are common at online travel agencies (e.g., Bid4Trips.com, SkyAuction.com) while other sites offer reverse auctions for the very flexible traveller (e.g., Priceline.com). Most major airlines provide online booking at their web sites but up to our best knowledge none of them offer automated online auctions (even though more than half of the tickets booked online are purchased through these sites). We believe that customer traffic at the sites of major airlines may generate enough demand so that excess seats on different flights can be pooled together and traded within the same auction (rather than trading pre-specified tickets one by one as the above mentioned travel agencies do). This allows customers to express their preferences better (with more flexibility than at Priceline.com) and the airlines to achieve a more efficient allocation of resources and thus more revenue.

This paper addresses the question of designing a simple iterative auction for trading excess seat capacity for an airline. Our study originated in the following real-life application where the airline was interested in auctioning off leftover seat capacity on flights *between two given cities*. The proposed format is an iterative sealed bid auction where bidders get instant feedback, including minimum bid suggestion for declined bids. Customers bid on round-trip tickets by specifying the number of tickets they wish to purchase and a per ticket bid amount. Bidders can express their preferences by submitting several flights, assumed to be equally acceptable, for both the inbound and the outbound leg. Winners are assigned all the seats they

*IBM Research Division, Thomas J. Watson Research Center, PO Box 704, Yorktown Heights, NY 10598. martaeso@us.ibm.com.

requested on one flight in each direction. Each bid is evaluated instantly: it is either rejected or provisionally accepted. Provisionally accepted bids may be displaced during the course of the auction by other bids, in which case the bidder is immediately notified. Note that bids can be displaced indirectly; that is, displaced bids do not even have to contain any of the flights that the new bid displacing them does. [1] A minimum bid suggestion is provided for rejected and displaced bids; these bids may be resubmitted with a higher bid amount, otherwise they are not considered again. The auction lasts for a given amount of time or until inactivity, whichever is later. Provisionally accepted bids become winners at closing, the owners of these bids are notified of their allocation and pay what they have bid.

Every iteration of the problem presented above can be modelled as a multi-unit combinatorial auction where the *bundles* customers bid on consist of the required number of tickets on acceptable flight pairs. With one bid a customer specifies an entire set of bundles which can be obtained by enumerating all the acceptable flight pairs. At most one of these bundles can be selected for a bidder; this requirement can be expressed as an exclusive OR of the bundles. Combinatorial auctions received much attention lately, see [3] for a survey, [5] for different bidding languages and [6] for background and computational complexity. Determining the winners (computing the best allocation of resources to the bidders) in a multi-unit combinatorial auction is NP-hard since the problem can be formulated as a multi-knapsack problem [4]. In [1] the authors demonstrate that commercial Mixed Integer Programming solvers are powerful and competitive compared to other heuristics for solving some classes of randomly generated (single-unit) combinatorial auction problems.

The above problem could be stated in a more general way. The airline could offer its leftover seat capacity on any number of flights, not just between two given cities. (Other products, like hotel rooms and rental cars could also be included in the bundles.) Customers could bid on arbitrary bundles of flights (not just pairs) and they could express a complex preference ordering on these flight combinations (rather than valuing all acceptable flight pairs the same). These generalizations can be incorporated into the Integer Programming model without making the winner determination problem computationally any more difficult (having a complex preference ordering might even help to avoid degeneracy of the multi-knapsack problem since these preferences translate to non-uniform objective coefficients). The remainder of the paper will mainly focus on the original problem setting but we will refer back to the extensions discussed in this paragraph from time to time.

---

[1] Assume that there are two flights in each direction (in1, in2, out1 and out2) with one leftover seat on each. The first bid is {in1, in2, out1, out2, \$150} and the second is {in2, out2, \$100}. Thus the provisional allocation is {in1, out1} and {in2, out2} for the two bidders, with a total profit of \$250. A third bid {in1, out1, \$200} arrives and displaces the second bid that does not have any common flights with the new bid.

The proposed mechanism is iterative: after a bid is submitted the winner determination problem is solved and feedback is provided in the form of a minimum bid suggestion for rejected bids. In general, bid submission, solving for the best allocation and providing some feedback for bid reformulation are repeated in a loop (until termination) in an iterative auction. The auction mechanism introduced above could be improved upon in all three components of this loop. First, bids are captured in our problem in the form of flights that are equally acceptable for the bidder. More sophisticated bid descriptions that capture customer preferences better (as indicated above) can be included as long as all the possible flight combinations along with their valuations can be efficiently enumerated. (For instance, limits on the time spent away from the departure city could be easily incorporated.) Second, since winner determination is the computationally intensive part of the loop it would make sense to execute it less often. That is, the allocation problem could be solved for batches of new bids that are collected over a given (short) period of time. This can result in more profit for the airline (since bids that are rejected when considered one by one cannot be used in a later iteration) and in more consistent bid reformulation signals (the minimum bid suggestion will fluctuate less, see Section 2.3). On the other hand it will take longer for the bidders to receive a feedback. Third, feedback for bid reformulation could be provided in a more general form. It is well known that no equilibrium item prices (i.e., prices for seats on individual flights) exist in a combinatorial auction if bidders have complementarities in their preferences (see e.g. [7]). However, suggestions could include extended flight sets that subsume those in the bid (which flights to include in the extended set requires business intelligence tools that are beyond the scope of this paper).

As a practical matter it is desirable for the auction to be simple (the rules are transparent and easy to understand), difficult to manipulate (by a single participant or by collusion) and incentive compatible (bidders reveal their true valuations). We believe the first two to be true of our design. Difficulty of manipulation is supported by the facts that bidders in this auction have very limited knowledge of each other and that the bids are sealed thus it is difficult to monitor compliance among ring members. However, we cannot prove incentive compatibility. While there might be a benefit to bid shading, because of the limited information bidders have about each other, we do not believe this poses an important problem. To further investigate these economic properties we would need to model bidder behavior which is outside of the scope of this paper.

The main contributions of this paper include the idea of computing minimum bid suggestions, a simple but flexible IP model and the use of heuristics (some of which are also IP based). In what follows we will first describe the roles of the participants in the auction, then discuss the winner determination problem in detail. We provide an Integer Programming formulation and outline heuristics and exact methods for solving it. Finally we present some computational results.

2. Participants of the auction. There are three roles in the auction, the auction organizer, the seller and the bidders. The airline plays the first two roles and the customers the third. The two roles of the airline should be treated separately since it might be advantageous to have an independent auctioneer. Having an independent auctioneer could increase the bidders' trust in the auction process since the identity, number and payment information of the bidders need not be disclosed to the airline. Also, an independent auctioneer will be able to handle offers from more than one airline.

2.1. The seller. Before the auction starts the seller needs to specify for the auctioneer the items for sale and any requirements that might restrict the feasibility of allocations. The airline needs to identify seat capacity and reserve prices (price below which the airline does not sell a seat) for the flights to be offered at the auction. In our setting reserve prices are the same for all flights but they could be differentiated based on historical demand information (like higher reserve prices for Friday night flights). Requirements on the feasibility of allocations could include limits on the number of winners, or, on the proportion of winners requiring at least a certain number of tickets. Also, a limit could be placed on the number of displaced bids when a new bid is accepted.

2.2. The bidders. Interested customers visit the auction site and browse through the schedule of flights available for auction. If they would like to participate then payment information (e.g., credit card) has to be submitted since winning bids are binding. After specifying some initial information (departure and destination airports, approximate dates for travel) bidders could be presented with a page like the one sketched out on Figure 1.a where they could fine tune their flight selection. The number of tickets requested and the bid amount (optional since minimum bid suggestion is provided otherwise) are also entered through this page. The system will immediately return with a page informing the bidder of acceptance or rejection (see Figure 1.b for a possible rejection page).

If a bid is rejected the customer may choose to increase the bid amount or cancel the bid. Cancelled bids are never used again by the system. If the bid is provisionally accepted the bidder will learn only this fact but not the exact flights that he is provisionally allocated. If a bid is later displaced the auction notifies the bidder (for instance, via e-mail) and also supplies a minimum bid amount needed for re-acceptance (assuming other bidders don't change their bids). Similarly as for the rejected bids, the bidder can either increase the bid amount or cancel the bid. Note that the bidder might want to change other factors of his bid than the bid amount (after a rejection or displacement). In the current setup this can be accomplished by cancelling the bid and submitting a new one, but any part of the bid could be modified in general (the minimum bid suggestion does not provide enough information for this, however).

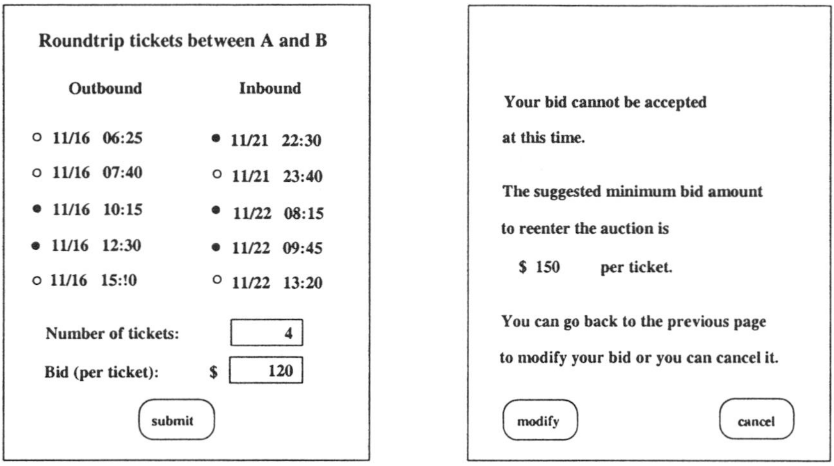

FIG. 1. *a. Page for entering a bid. b. Rejection page with minimum bid suggestion.*

We assume that bidders have private valuations for the goods and that they behave rationally. Also, if their bid is rejected or displaced bidders resubmit the bid as long as the suggested minimum bid amount is below their valuation.

2.3. The auctioneer. The main tasks of the auction organizer are to collect information from the seller, open the auction, collect the bids and respond to them, close the auction and announce the winners. The closely related problems of determining whether a bid is accepted or rejected and computing the minimum bid amount are the computationally challenging operations.

While the auction is open a list of provisionally accepted bids along with the corresponding seat assignment is maintained by the auctioneer. When a new bid arrives it is evaluated and is either accepted or rejected. Evaluation (also called winner determination) means computing the "best" allocation (which maximizes the seller's utility) given the provisionally accepted bids and the new bid. (Comparing allocations and solving the winner determination problem will be discussed in Section 3.) If the new best allocation is better than the best allocation from the previous iteration then the new bid is accepted and the best allocation is updated to be the new allocation. Otherwise the bid is rejected and the best allocation remains the same.

To compute the minimum bid suggestion we have to determine an amount for the new bid at which the new best allocation becomes preferable over the previous one. This can be accomplished by pretending that the new bid is accepted at reserve prices and computing the new best allocation (which is computationally as complex as solving the original winner determination problem). If none of the provisionally accepted bids is displaced

flights	seats	Bid 1	Bid 2	Bid 3		Bid 4	Bid 2	Bid 3
out 1	2	2		1		1		1
out 2	1		1		Bid 4		1	
in 1	2	2			replaces	1		
in 2	1		1	1	Bid 1		1	1
		$105	$100	??		$220	$100	??
compensate				$310				$100
overbid				$105				$100

FIG. 2. *Example of fluctuating minimum bid amounts.*

(the allocation can be different however) then the minimum bid amount needed to enter the auction is the sum of reserve prices on the assigned flights. Otherwise there are displaced bids and the bidder must compensate the seller for the lost revenue or pay the reserve price, whichever is higher. We propose two different methods for compensating the seller: pay the full amount needed for the new best allocation to overtake the previous allocation, or pay a little more (per ticket) than the highest payment from the displaced bids.

The first method of fully compensating the seller has the advantage that successive allocations become better and better for the seller at each iteration. However, the minimum bid amounts do not necessarily increase monotonically which might be counterintuitive for bidders familiar with single-item ascending price auctions. The example of Figure 2 illustrates that fluctuations may happen with both of the minimum bid computation rules. For the example we assume that there are no reserve prices and that the seller wants to maximize the revenue collected from the bidders. There are two outbound and two inbound flights with corresponding seat availabilities. When Bid 3 arrives the first time it has to displace both Bid 1 and Bid 2 to be accepted. Thus the bidder would need to pay 2*$105 + $100 = $310 (plus a bid increment of say $1) to get accepted. Assume that this is above the bidder's valuation and he stays away. Now Bid 4 arrives and replaces Bid 1, leaving one seat on both out 1 and in 1 unallocated. Should the new bidder resubmit his bid now, only Bid 2 would need to be displaced and the bidder pays $100 (+$1).

With the second method of (per ticket) overbidding the highest displaced bid the minimum bid suggestion may still fluctuate; in the above example Bidder 3 would need to pay $105 (+$1) in the first case and $100 (+$1) in the second case. In addition to this the seller's revenue might not be monotone (he collects only $105 (+$1) compared to $310 if the new bid offers at least the minimum bid amount). This method may also leave more unallocated seats at the end of the auction.

As we have mentioned in the Introduction, solving the winner determination problem for batches of new bids will result in less fluctuation in the minimum bid amounts (for the full compensation case). Indeed, optimizing with more than one new bid at once rather than incrementally provides better allocations with more of the resources assigned. Several new bids together may displace a set of accepted bids, sharing the cost of compensation. Note that even if bids are evaluated in batches, the minimum bid suggestion is computed for each bid one by one, assuming that the other bids remain the same. Thus some bid amount needs to be known for all new bids. While reserve prices may play this role the resulting minimum bids would be over-estimated in this case.

We have experimented with both minimum bid computation rules (see Section 3.4) and arrived to the conclusion that the full compensation method is superior if bidders can accept fluctuating minimum bids.

3. The winner determination problem. In this section we will focus on how to find the best allocation given the set of provisionally accepted bids and the new bid(s). We discuss how to compare different allocations, formulate the winner determination problem as an Integer Program and outline heuristics and exact methods for solving it. Computational results are also presented at the end of this section.

3.1. Comparing allocations. The first question to ask is that given a set of bids and seat (resource) availability, what objective should the seller use to compare different allocations. These objectives could be the seller's profit (the total amount collected from the bidders minus the reserve prices of the allocated seats), the number of accepted bids or the number of seats allocated. The profit of the seller is the best base for comparison since otherwise the auction could be easily manipulated. (If the allocation with the higher number of accepted bids is preferred then bidders wishing to purchase multiple tickets would have a better chance of winning by submitting multiple bids for single tickets instead of one bid for multiple tickets. If the allocation with the more seats assigned is preferred then in case of high enough demand no new bids are accepted after all the seats are provisionally allocated.) An allocation is optimal if it is best with respect to the chosen objective for the given the set of bids. Other objectives than profit maximization could be incorporated as part of a composite objective or as side constraints. We used profit maximization as the seller's objective in our computational experiments.

3.2. IP formulation of winner determination. We can formulate the Winner Determination problem as an Integer Program. Let us introduce some notation first. The flights available for auction are given as $S = \{1, \ldots, m\}$ with corresponding seat availability (capacity) $c_1, \ldots, c_m$ and reserve prices $r_1, \ldots, r_m$. The flights are indexed by j. There are n bidders, bid i consists of the set of acceptable flights in both directions $E_i \subset S$ and $F_i \subset S$, the number of tickets requested t_i and the per ticket bid amount b_i. Let us enumerate the acceptable flight pairs and index them with $k = 1, \ldots, k_i$ where $k_i = |E_i||F_i|$. If the bid is provisionally accepted then the bidder will be allocated t_i seats on one flight in E_i and one flight in F_i and will pay $t_i b_i$ if the bid is a winner.

To formulate the winner determination problem we introduce binary variables for each acceptable flight pair of each bidder indicating whether the flight pair is chosen in the solution or not:

$$x_{i,k} = 1 \text{ if flight pair } k \text{ is chosen for bidder } i; k = 1, \ldots, k_i.$$

There are two sets of constraints: the seat availability cannot be exceeded for any of the flights and at most one of the flight pairs can be chosen for any bidder. Introduce $a^j_{i,k}$ to denote the number of tickets on flight j for flight pair k of bidder i (that is, $a^j_{i,k}$ is t_i if j is one of the two flights in the pair and 0 otherwise). With this notation the resource limit constraints can be written as

$$\sum_{i=1}^{n} \sum_{k=1}^{k_i} a^j_{i,k} x_{i,k} \leq c_j, \quad j = 1, \ldots, m;$$

and the second set of constraints (choose at most one flight combination for each bidder) as

$$\sum_{k=1}^{k_i} x_{i,k} \leq 1, \quad i = 1, \ldots, n.$$

We also include the integrality restrictions on the variables:

$$x_{i,k} \in \{0, 1\}, \quad i = 1, \ldots, n; \quad k = 1, \ldots, k_i.$$

The objective is to maximize the seller's profit (payment collected from winners minus reserve prices):

$$\sum_{i=1}^{n} \sum_{k=1}^{k_i} \left(t_i b_i - \sum_{j=1}^{m} a^j_{i,k} r_j \right) x_{i,k}.$$

We provide a sketch of the formulation in Figure 3 for a simple case where the reserve prices are assumed to be constant (so they can be omitted from the objective function). Columns of the matrix correspond to the acceptable flight pairs. Note that the flights can be divided into two sets, the outbound and inbound flights (as viewed from say the first of the two cities). Each flight pair will have one entry in each set, the number of tickets required by the bidder. The sketch makes it clear that this is a multi-unit combinatorial auction where the bids are for bundles of flights (we can introduce a fictitious flight for each bidder to account for the second set of constraints.) Observe that the problem matrix has $m + n$ rows, $\sum_{i=1}^{n} k_i$ columns and three nonzeros per column (thus the problem matrix is very sparse).

	bid 1						bid 2				bid 3									
	40		.....			40	20	.....		20	30			.....				30		
out 1	4	4	4								2	2							<=	4
out 2							1	1					2	2					<=	4
out 3				4	4	4			1	1					2	2			<=	4
out 4																	2	2	<=	4
in 1	4			4							2		2		2		2		<=	5
in 2		4			4		1		1										<=	5
in 3			4			4		1		1		2		2		2		2	<=	6
bid 1	1	1	1	1	1	1													<=	1
bid 2							1	1	1	1									<=	1
bid 3											1	1	1	1	1	1	1	1	<=	1

FIG. 3. *Integer Programming formulation of the winner determination problem.*

As we have mentioned in the Introduction, any set of flights could be offered for auction. In this case the set of rows could be subdivided so that flights in the same subset have the same origin and destination. Also, customers could bid on general bundles, probably containing at most one flight from each subset. Finally, complex preference orderings on the acceptable flight combinations could be incorporated into the objective function.

To enforce that a certain bid is accepted (which is needed for minimum bid computation, see Section 2.3) exactly one of the flight pairs must be chosen for the bid. This can be achieved by setting the constraint in the second set corresponding to this bidder to equality:

$$\sum_{k=1}^{k_i} x_{i,k} = 1 \quad \text{to accept bid } i.$$

Restrictions imposed by the seller (mentioned in Section 2.1) can be added in the form of additional (side) constraints. For instance, limits on the total number of accepted bids can be expressed as

$$L \leq \sum_{i=1}^{n} \sum_{k=1}^{k_i} x_{i,k} \leq U.$$

Note that flight pairs with non-positive objective coefficients (i.e., reserve price exceeds bid amount) will never participate in any optimal allocation. Therefore, columns (and variables) corresponding to such flight pairs can be simply left out of the formulation before solving the problem.

3.3. Solving the winner determination problem. Here we will outline heuristic and exact solution methods for the winner determination problem. We will discuss how to solve the problem for one new bid or a batch of new bids, and how to use the same algorithm to compute the minimum bid suggestion for one rejected/displaced bid. The drawback of the first two heuristics described below is that they cannot efficiently handle the winner determination problem for a batch of new bids.

Another difficulty that may arise when using heuristics is that small changes in the formulation might require the development of different specialized algorithms. This is however not the case with the heuristics discussed here.

We will present the heuristics and the exact methods in increasing order of their computational complexity.

3.3.1. Obvious inclusion. This is a very straightforward heuristic. Given an already existing allocation the new bid is accepted only if there is enough capacity to accommodate it without reallocating the already (provisionally) accepted bids. This happens when there is enough unallocated seating capacity for both segments of an acceptable flight pair. We call this the *obvious inclusion* of the new bid. If there is enough capacity on more than one acceptable flight pair then one has to be chosen based on some rule (e.g., choose the flight pair with the smallest reserve price, or choose at random). Side constraints can be incorporated by rejecting the new bid if its acceptance would violate the constraint.

If the winner determination problem is solved for a batch of new bids then order the bids (e.g., randomly, or in the order of their arrival) and accept/reject them one by one using obvious inclusion. This procedure could be repeated for different orderings of the new bids (if there are few new bids, all orderings could be considered, otherwise repeat only a constant number of times or until some time limit is reached).

Since the existing allocation cannot be modified by definition of this method, minimum bid suggestions are made only if there is enough unallocated capacity for the bid but the bid amount does not reach the reserve price.

We check for obvious inclusion *before* any other method since it is computationally very cheap. However, our experiments confirm that both the seller's profit and the allocated capacity is weak if this heuristic is used alone.

3.3.2. Simple greedy heuristic. This heuristic uses obvious inclusion repeatedly. We consider the already accepted bids and the new bid(s) together, order them randomly and accept/reject them one by one using obvious inclusion. This procedure is repeated a constant number of times or until a time limit is reached. If the overall best allocation is better than the allocation with the previously accepted bids then the new allocation replaces the old one.

To compute the minimum bid suggestion for a rejected or displaced bid consider only those orderings of this bid and the bids in the new allocation where this bid is the first. Again, a sequence of obvious inclusions is applied to these orderings and the overall best solution is used for minimum bid computation.

This heuristic is computationally inexpensive and results in a better quality solution than the previous one.

3.3.3. IP based heuristic. To compute the new best allocation the winner determination IP of Section 3.2 needs to be solved for the already accepted bids and the new bid(s). The new allocation replaces the old one if it is better. To compute a minimum bid suggestion the IP is considered with the bids in the new allocation and the rejected/displaced bid (which is forced into the solution by setting its constraint to equality).

Solving the winner determination IP to optimality might not be a practical approach if the size of the input is too large. During a typical IP solution process (like Branch-and-Bound) linear relaxations of the IP have to be solved repeatedly. Since the computational effort to solve the linear relaxations is determined by the row dimension of the problem matrix, our IP based heuristic uses only a submatrix of the original formulation. In particular, the heuristic considers only those flights (rows) that are acceptable for bidders who are currently assigned seats on flights acceptable for the new bid. We achieve this by "fixing" the allocation to other bids that currently do not have any seats assigned to them on flights the new bidder would accept. The fixing is accomplished by setting the variables corresponding to the flight assignments to one. Note that considering only the above flights indeed restricts the problem since the new bid could indirectly displace some other bids that have no common flights with the new bid (as we have illustrated in the Introduction).

Any side constraint that was included in the original IP formulation can also be added to the restricted IP. This heuristic is still computationally intractable (NP-hard) in theory since it is still a multi-knapsack problem. However, the size of the problem is expected to be much smaller than the size of the original formulation and thus we expect to obtain a solution faster. Any of the methods sketched out in the next section could be applied.

3.3.4. Exact solution of the IP. If the problem size is reasonable, commercial IP packages (which employ a Branch-and-Bound backbone) can be used to solve the Integer Program to optimality. This is what we did in our experiments. However, it is likely that more sophisticated methods have to be devised especially if the set of flights offered for auction is extended or if the the number of flights or the flight capacities are large.

A generic IP solver could be significantly enhanced. Using problem specific information a custom code could be built that uses, for instance, column generation, better branching rules and warmstarting. In a column generation scheme (also called pricing) we start with a small, promising subset of the variables and add other variables to the formulation iteratively. Our implementation is under way using BCP, an LP formulation based parallel Branch-and-Cut-and-Price framework for Mixed Integer Programs. BCP is a module in COIN-OR, an Open Source initiative for the OR community (see [2]). Special branching rules can be devised that take the problem structure into consideration. For instance, instead of branching on a variable (which assigns a flight pair to a bidder) we could branch on whether a particular flight leg is assigned to a bidder or not. Variables corresponding to the same bidder form SOS Type 1 sets which can also be branched on (although our computational results did not improve when we specified these sets for the solver engine). Warmstarting is a method to use information from a previously solved problem to speed up computations in the current one. Since the subsequent IPs differ only in a few bids the search tree of the previous optimization is likely to be a good starting point for the current problem.

3.4. Some computational results. Since no real data was available to us, we have used uniformly generated data in our tests. First we constructed the flight availability information (m, c_j, r_j) by hand, with the seat capacity uniformly distributed among the flights. The number of flight segments that could be chosen in both directions $(|E_i|, |F_i|)$ and the number of tickets (t_i) were both uniformly distributed between one and four. Per ticket bid amounts (b_i) were chosen uniformly from the interval [\$200, \$300] The number of bids (n) to generate was chosen so that the total demand is twice or four times the supply (the first case representing low demand while the second high demand). The bids were taken in the order they were generated and either provisionally accepted or rejected. Rejected and displaced bids were never resubmitted in our tests. Reserve prices were set to \$200 for all the flights and we assumed a minimum bid increment of \$1. We did not try to model agent behavior with the uniformly generated problem instances, our only goal was to test the viability of the heuristics and the Integer Programming based solution methods.

We have implemented the three heuristics and the exact method for solving the winner determination problem, as discussed above. We solved the IPs using IBM OSL, a commercial solver package. Both minimum bid computation rules were implemented (fully compensate the seller and overbid the highest displaced bid). In half of the experiments we included a side constraint that limits the number of displaced bids to two in each iteration. Our tests were carried out on an IBM RS6000 43P Model 40 computer.

Figures 4 and 5 summarize our results for a small example. There are 50 seats available, 40 bids were generated for the low demand and 80 bids for the high demand case. The tables contain the allocated capacity, the seller's profit and the running time for the three heuristics, for both mini-

	compensate no limit on displ			overbid no limit on displ		
	alloc cap	profit	time	alloc cap	profit	time
obvious	44.6	11.20	0	42.8	10.73	0
greedy	46.2	11.80	0.2	42.2	11.19	0.1
IP heur	48.8	12.66	6.8	46.6	12.27	8.6

	compensate displ ≤ 2			overbid displ ≤ 2		
	alloc cap	profit	time	alloc cap	profit	time
obvious	43.2	11.21	0	42.2	10.45	0
greedy	46.4	12.06	0.1	41.8	10.97	0.1
IP heur	49.0	13.07	6.1	48.6	12.69	5.7

FIG. 4. *Experiments: 50 seats, 40 bids (low demand).*

	compensate no limit on displ			overbid no limit on displ		
	alloc cap	profit	time	alloc cap	profit	time
obvious	48.2	12.02	0	48.2	12.22	0
greedy	48.8	12.65	0.4	45	12.55	0.3
IP heur	50	13.66	32.9	49.4	13.82	17.1

	compensate displ ≤ 2			overbid displ ≤ 2		
	alloc cap	profit	time	alloc cap	profit	time
obvious	49	12.60	0	48.8	12.23	0
greedy	48.8	12.76	0.3	46.2	12.59	0.3
IP heur	50	13.62	38.6	49.6	13.75	19.3

FIG. 5. *Experiments: 50 seats, 80 bids (high demand).*

mum bid computation rules and with or without the limit on the number of displaced bids. That is, there are altogether 12 experiments for each problem instance. The table contains averages for five instances.

Using our synthetic data we can conclude that most of the seats can be allocated at the end of the auction in the high demand case no matter which heuristic we use. However, the seller's profit improves dramatically with the more sophisticated heuristics. Obviously, running times are the exact opposite. Observe also that the allocated capacity is higher with the "compensate the seller" minimum bid computation rule than with the "overbid the highest displaced bid" rule. Profits are also significantly higher with the first rule in the low demand case. Also note that limiting the number of displaced bids does not hurt the allocated capacity or profits in the high demand case.

4. Conclusions. We have presented an iterative design for airline seat auctions which we believe could be profitable to introduce in the travel industry, since it is easy to understand and is trustable for consumers and it can be implemented using readily available optimization software. We introduced a simple but flexible Integer Programming model which was used both for winner determination and for computing minimum bid suggestions, and also experimented with heuristics and exact methods for solving the model. In the future we plan to further investigate the economic properties of the design and the computational feasibility of more complex settings outlined in the paper.

Acknowledgements. Thanks to Manoj Kumar for posing the original problem and to Peter Eso, David Parkes, Laszlo Ladanyi and the anonymous reviewers for providing valuable suggestions.

REFERENCES

[1] A. Andersson, M. Tenhunen, and F. Ygge, *Integer Programming for Combinatorial Auction Winner Determination*, Proc. of the Fourth International Conference on Multiagent Systems (ICMAS-00), 2000.

[2] *COIN-OR: Common Optimization Interface for Operations Research*, http://www.coin-or.org.

[3] S. de Vries and R. Vohra, *Combinatorial Auctions: A Survey.* Technical report, MEDS, Kellogg Graduate School of Management, Nothwestern University. 2000.

[4] M.R. Garey and D.S. Johnson, *Computers and Intractability.* W.H. Freeman, 1979.

[5] N. Nisan, *Bidding and Allocation in Combinatorial Auctions*, Manuscript, available at http://www.cs.huji.ac.il/~noam/mkts.html.

[6] M.H. Rothkopf, A. Pekeč, and R.M. Harstad, *Computationally Manageable Combinatorial Auctions.* Management Science, 44(8): 1131–1147, 1998.

[7] M.P. Wellman, W.E. Walsh, P.R. Wurman, and J.K. MacKie-Mason, *Auction Protocols for Decentralized Scheduling. Games and Economic Behavior*, 35(1/2): 271–303, April/May 2001.

AN INTEGER PROGRAMMING FORMULATION OF THE BID EVALUATION PROBLEM FOR COORDINATED TASKS*

JOHN COLLINS AND MARIA GINI†

Abstract. We extend the IP models proposed by Nisan and Andersson for winner determination in combinatorial auctions, to the problem of evaluating bids for coordinated task sets. This requires relaxing the free disposal assumption, and encoding temporal constraints in the model. We present a basic model, along with an improved model that dramatically reduces the number of rows by preprocessing the temporal constraints into compatibility constraints. Experimental results show how the models perform and scale, and how they compare with a stochastic solver based on Simulated Annealing. On average, the IP approach finds optimum solutions significantly faster than Simulated Annealing, but with an extreme level of variability that may make it impractical in time-constrained agent negotiation scenarios.

Key words. automated negotiation, multiattribute combinatorial auctions, Integer Programming.

AMS(MOS) subject classifications. 90C10, 68T20.

1. Introduction. Business-to-business e-commerce is expanding rapidly, letting companies both broaden their customer base and increase their pool of potential suppliers. Negotiating supplier contracts for the multiple components that often make up a single product is complicated because time dependencies introduce a significant scheduling risk. Current e-commerce systems typically rely on either fixed-price catalogues or auctions [9], and they often don't deal effectively with the time dimension.

The University of Minnesota MAGNET (Multi-Agent Negotiation testbed) [3] system is designed to support the negotiation of contracts for coordinated tasks among a population of independent and self-interested agents. Dealing with coordinated tasks adds some major complexities to the auction model of agent interaction. First, because tasks have a temporal duration and precedence constraints, any bid selection algorithm must insure the temporal feasibility of the bids accepted. Second, because each of the tasks in the set of coordinated tasks is necessary to achieve the overall goal, any bid selection algorithm must insure complete coverage of the tasks. Third, delays and failures that might occur during the execution of the contracted tasks can threaten the accomplishment of the overall goal. This introduces the need to assess scheduling risk and to account for risk in the bid selection algorithm.

---

*Work supported in part by the National Science Foundation, awards NSF/IIS-0084202 and NSF/EIA-9986042.

†Department of Computer Science and Engineering, University of Minnesota, Minneapolis, MN 55455. {jcollins,gini}@cs.umn.edu.

Several papers have been published recently that deal with linear and integer programming formulations of the allocation problem in combinatorial auctions [1, 10]. This leads the the obvious question: can the MAGNET bid-evaluation problem be formulated as a linear or integer programming problem. If so, there are well-known and reasonably efficient evaluation procedures for these problems, although in general the integer programming problem is NP-complete. We show that an integer programming approach appears to be a feasible approach to the MAGNET bid-evaluation problem, and we evaluate its performance and suitability for a time-constrained agent reasoning process.

In the next section, we give a simple example that will help understand the details of our approach. Section 3 presents a straightforward IP model of the MAGNET bid-evaluation problem, and follows up with an improved version that uses preprocessing to generate a more compact representation. In Section 4, we present experimental results comparing the two IP formulations and a heuristic search method based on Simulated Annealing, and we show why the Simulated Annealing approach might be valuable, even though its average performance is much worse than the improved IP formulation. Finally, Section 5 relates this work to earlier work in solving combinatorial auction problems, and Section 6 presents our conclusions and suggestions for further work in this area.

2. Example. Suppose we have a job to do that involves performance of a set of coordinated tasks within a limited time frame, and we wish to minimize the cost of the job. Examples might include constructing a building or a bridge, evacuating an island [12], shipping a large piece of industrial equipment overseas, or establishing a multi-link communication channel. The resources needed to perform those tasks must be acquired from self-interested suppliers, who are attempting to maximize the value of the resources under their control. It is the job of a MAGNET Customer agent to use an auction process to obtain a set of commitments for those resources, that can be composed into a temporally feasible plan, at a minimum price.

Figure 1 shows an example task network. It is a directed acyclic graph, with arcs representing precedence relations. (In this paper, the notion of "precedence" means that if task A precedes task B, written as $A \prec B$, then task A must be completed before task B can start. No delay is implied between the end of task A and the start of task B.) The numbers in parentheses are the expected durations of each task. We expect task duration data, as well as data on duration variability, resource availability, and supplier reliability, to be collected and made available from the MAGNET market infrastructure [3].

Suppliers submit bids on sets of tasks based on their own resource availabilities and costs, and based on a Request for Quotes (RFQ) submitted by a potential customer. A bid includes a set of tasks and a price, along

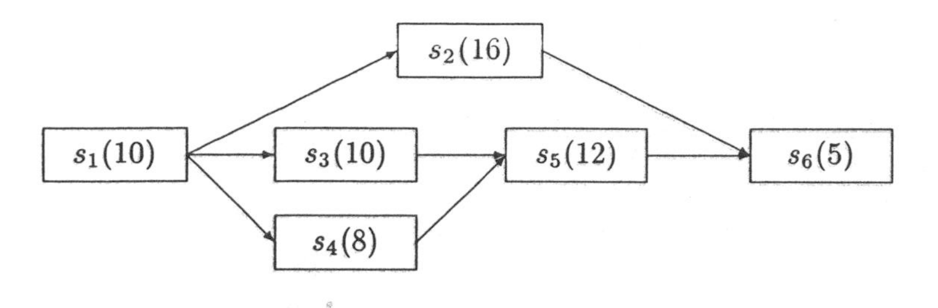

FIG. 1. *Example task network.*

with timing data, including duration and the earliest and latest times the task(s) may be started. When composing the RFQ for a plan, the customer must give suppliers some guidance about when the work must be done. This is done by specifying a *time window* for each task that gives the earliest start time and latest finish time. In generating this schedule, the customer has two conflicting goals:

1. Ensure that the bids received can be composed into a feasible plan. This can be done by specifying time windows that do not overlap.
2. Specify relatively wide, and possibly overlapping time windows, in hopes of attracting more bids and lower prices. The risk in doing this is that many bid combinations will not compose feasibly because their task time windows will be in conflict.

If we take the first approach, then the bid-evaluation process reduces to a variation on the combinatorial-auction winner-determination problem [16], without the free-disposal assumption[1]. The second approach leads to a much more difficult and interesting bid-evaluation problem, which is the subject of this paper.

The plan in Figure 1 has a *makespan* of 37 time units. This is the longest path through the graph. If we have a deadline for plan completion of 40 units, then we have an *overall slack* of 3 units, or about 8%.

Table 1 shows some bids that might be received for this plan. Several observations are apparent:

- Each bid may specify a "bundle" of tasks. A single price is given for the bundle.
- In this example and in the remainder of this paper, we assume that bids give early start, late start, and duration data for each task individually. This assumption can be relaxed.
- Bids b_1 and b_3 cannot both be accepted, because they both specify task s_5. Each task must be allocated to exactly one bid.
- Bids b_1 and b_2 cannot both be accepted, because the precedence relation $s_5 \prec s_6$ would be violated. This because the earliest time

[1]The free-disposal assumption states that items can remain unallocated without penalty.

TABLE 1
Example Bids.

Bid	Task	Price	Early Start	Late Start	Duration
b_1	s_1	200	1.5	3.0	11.0
	s_3		12.5	14.0	8.5
	s_5		22.0	23.0	13.0
b_2	s_2	290	10.0	16.0	18.0
	s_6		28.0	33.0	6.0
b_3	s_4	160	9.0	13.0	6.0
	s_5		15.0	19.0	14.0
b_4	s_4	20	10.0	15.0	11.0

task s_5 could be completed under bid b_1 is 35, while the latest time task s_6 could start under bid b_2 is 33.

- Bids b_1 and b_4 cannot both be accepted. This is more subtle. The precedence relations $s_1 \prec s_4$ and $s_4 \prec s_5$ can be satisfied individually. However, when we attempt to combine the two bids, we see that, in order to satisfy $s_1 \prec s_4$, the early finish time of task s_4 is pushed back to 23.5, and $s_4 \prec s_5$ is violated.

We'll use this example to examine details of our bid-evaluation process.

3. IP formulations. We start by introducing some notation. A plan consists of a set $\mathcal{S}$ of tasks $s_j, j = 1..m$. Each task s_j has a precedence set $\mathcal{P}_j = \{s_{j'} | s_{j'} \prec s_j\}$, the set of tasks $s_{j'}$ that must be completed before s_j is started. At the conclusion of some bidding process, we have a set $\mathcal{B}$ of bids $b_i, i = 1..n$. Each bid b_i specifies a set of tasks $\mathcal{S}_i$ and a price p_i. For each task s_j, a bid b_i that includes the task ($s_j \in \mathcal{S}_i$) may specify an early start time e^i_j, a late start time f^i_j, and a duration d^i_j.

We have also defined a risk factor r_i, associated with each bid, that is based on static factors, such as the reputation of the bidder. The derivation of r_i is outside the scope of this discussion, we include it for completeness.

3.1. A straightforward model. First, we present a "direct" model, in which the coverage and feasibility constraints are directly represented. We associate a 0/1 variable x_i with each bid b_i, with the sense that in a solution where $x_i = 1$, b_i is accepted. No pre-processing is required other than composing the constraint rows. If we include only static risk factors (those that are determined strictly by the set of bids chosen, and not by scheduling considerations), the formulation of the basic bid-evaluation problem is:

Minimize:

$$\sum_{i=1}^{n} (p_i + r_i) x_i.$$

Subject to:

- Bid selection – each bid is either selected or not selected. These are the integer variables that make this an integer programming problem.

$$x_i \in \{0, 1\}.$$

- Coverage – each task s_j must be included exactly once.

$$\forall j = 1..m \sum_{i | s_j \in \mathcal{S}_i} x_i = 1.$$

Note that under the free disposal assumption, each task would be included at most once, rather than exactly once.

- Local feasibility – each task s_j must be able to start after the earliest possible completion time of each of its predecessors $s_{j'}$. This constraint ignores global feasibility. In other words, here we are looking only at the start times of a particular task and its immediate predecessors.

$$\begin{aligned}&\forall j = 1..m, \forall i | s_j \in \mathcal{S}_i, \forall i' | s_{j'} \in (\mathcal{S}_{i'} \cap \mathcal{P}_j),\\ &x_i f_j^i \geq x_{i'}(e_{j'}^{i'} + d_{j'}^{i'}) - M(1 - x_i)\end{aligned}$$

where M is a "large" number (we typically use a value of 10^{12}), and the last term $M(1 - x_i)$ is used to make the constraint satisfied in the case where $x_i = 0$.

In our example, the precedence relations between b_1 and b_4 would be expressed as

$$\begin{aligned}x_4(15.0) &\geq x_1(1.5 + 11.0) - M(1 - x_4)\\ x_1(23.0) &\geq x_4(10.0 + 11.0) - M(1 - x_1).\end{aligned}$$

- Global feasibility – each task must be able to start after the earliest possible completion time of each of its predecessors, where the predecessors may in turn be constrained not by their bids, but by their respective predecessors.

$$\begin{aligned}&\forall j = 1..m, \forall i | s_j \in \mathcal{S}_i, \forall i' | s_{j'} \in (\mathcal{S}_{i'} \cap \mathcal{P}_j), \forall i'' | s_{j''} \in (\mathcal{S}_{i''} \cap \mathcal{P}_{j'}),\\ &\quad x_i f_j^i \geq x_{i'} d_{j'}^{i'} + x_{i''}(e_{j''}^{i''} + d_{j''}^{i''}) - M(1 - x_i)\\ &\forall j = 1..m, \forall i | s_j \in \mathcal{S}_i, \forall i' | s_{j'} \in (\mathcal{S}_{i'} \cap \mathcal{P}_j),\\ &\quad \forall i'' | s_{j''} \in (\mathcal{S}_{i''} \cap \mathcal{P}_{j'}), \forall i''' | s_{j'''} \in (\mathcal{S}_{i'''} \cap \mathcal{P}_{j''}),\\ &\quad x_i f_j^i \geq x_{i'} d_{j'}^{i'} + x_{i''} d_{j''}^{i''} + x_{i'''}(e_{j'''}^{i'''} + d_{j'''}^{i'''}) - M(1 - x_i)\\ &\quad \vdots\end{aligned}$$

In our example, the infeasibility between b_1 and b_4 is captured by the constraint

$$x_1(23.0) \geq x_4(11.0) + x_1(1.5 + 11.0) - M(1 - x_1).$$

The number of constraints generated by these formulas is highly variable, depending strongly on the details of the submitted bids and how they interact with the precedence network in the plan. For example, in the 5 tasks – 20 bids case described in Section 4, the count varied from 10 to 24000.

3.2. Collapsing the feasibility constraints. The formulation given in Section 3.1 is correct, but it can be dramatically improved, based on several observations. The first is that we can pre-process the coverage constraints to reduce the number of bids. If there is any task s_j for which only one bid b_i has been received (we'll call bid b_i a "singleton" bid for task s_j), b_i must be part of any complete solution. Bids b_k that conflict with b_i can then be discarded. In more formal terms,

$$\forall j | \sum_{i | s_j \in \mathcal{S}_i} 1 = 1, x_i = 1, \forall k | \mathcal{S}_i \cap \mathcal{S}_k \neq \emptyset, x_k = 0.$$

This test is repeated until no further singleton bids are detected.

Next, we make the following observations regarding the feasibility constraints:

1. We have generated feasibility constraints between bids that cannot possibly be part of the same solution, because they contain overlapping task sets. The coverage constraints will ensure that only one bid will be chosen to cover each task. We can discard such constraints immediately. In our example, this means that we need not generate or evaluate any feasibility constraints that include both b_1 and b_3.
2. The feasibility constraints can be greatly simplified by doing the arithmetic during preprocessing, and including only those constraints that can have an impact on the outcome. This way, we eliminate all the feasibility constraints shown in Section 3.1, and replace them with a much smaller number of simple compatibility constraints. In other words, if we start with a constraint of the form

$$x_i f_j^i \geq x_{i'}(e_{j'}^{i'} + d_{j'}^{i'}) - M(1 - x_i),$$

we compute $f_j^i - (e_{j'}^{i'} + d_{j'}^{i'})$. Just in case the result is negative, we include a constraint of the form

$$x_i + x_{i'} \leq 1$$

which will prevent both x_i and $x_{i'}$ from being part of the same solution. This simplification can be similarly applied to the global feasibility constraints. In general, such constraints tell us that for some combination of n bids, at most $n-1$ of them may be part of a solution.

If either x_i or $x_{i'}$ in the above formula is a singleton, then clearly the other cannot be part of a solution, so it can be eliminated. Also, if both x_i and $x_{i'}$ are singletons, then we know the problem cannot be solved.

In our example, we can observe the infeasibility between b_1 and b_4 during pre-processing, and rather than generate the formula given above in Section 3.1, we replace it with

$$x_1 + x_4 \leq 1.$$

3. If we have successive tasks in the same bid, we can filter the bids themselves for internal feasibility prior to evaluation. After the previous step, any constraints of the form $x_i + x_i \leq 1$ represent bids x_i that can be discarded.

3.3. Minimizing completion time. If we want to minimize the time to complete the plan, we must develop an expression for the completion time. To begin with, we define t_0 as the (fixed) start time of the plan. Then we need to determine the latest time at which some task will be completed. We needn't consider all tasks, just the "leaf tasks," those that have no successors. If we ignore precedence constraints, the earliest possible completion time t_c for the plan as a whole is the maximum early finish time of any leaf task for a given bid assignment. Since in any valid bid assignment, only one bid is chosen for any given task, we can express this as

$$t_c = \max_{j | \forall k, s_j \notin \mathcal{P}_k} \sum_{i | s_j \in \mathcal{S}_i} x_i (e_j^i + d_j^i)$$

where a task s_j that has no successors is one that is in the predecessor set of no other tasks.

Unfortunately, it doesn't work to ignore precedence constraints. There may be a task in the precedence set of any given leaf task s_j whose early finish time in a given bid assignment will prevent s_j from starting at its early start time. To avoid this problem, while taking advantage of the precedence constraints we've already developed, we do two things. First, we must have a single task whose completion marks the end of the plan; to ensure that this is the case, we create a dummy task s_c, with 0 duration. We then define t_c as the start time of task s_c. We don't define a bid for this task, because we want to use its start time as a variable. We also define the plan start time t_0. Then we have to add a completion-time term to the objective function, which now reads:

Minimize:

$$W_c \sum_{i=1}^{n} (p_i + r_i)x_i + W_t(t_c - t_0)$$

where W_c is the relative weight given to cost, W_t is the relative weight given to completion time, and $(t_c - t_0)$ is the total makespan of the plan.

Next, we add an additional set of feasibility constraints, as given above in Section 3.1, to constrain the dummy task to start later than the completion times of all the leaf tasks. This set will include the local and global feasibility constraints, expanded recursively to the root tasks, substituting t_c for $x_i f_j^i$. Since t_c is a variable that appears in the objective function, its final value with be the earliest time that is greater than the maximum early completion time over all leaf tasks for any given bid assignment.

This approach does not work with the simplified form of the feasibility constraints as given in Section 3.2. There, we have discarded the temporal information in preprocessing, and are left with simple compatibility constraints. This deprives the IP solver of the information necessary to operate on completion time. An alternative approach is to make completion time be a constraint, rather than a factor in the objective function. This would typically require multiple passes through the IP solver to find an acceptable combination of price and completion time.

4. Experimental results. To illustrate the effectiveness of our formulation, we have implemented both the original formulation given in Section 3.1, and the pre-processed formulation given in Section 3.2, without the completion-time extension. Due to resource constraints, we were only able to run the original formulation on very small problems. We also ran our Simulated Annealing (SA) [13] search engine [4] on the same problems for comparison. In general, results show that our IP formulation is practical for moderate-sized problems, though it scales exponentially. On average, it performs better than the SA approach (as it was tuned for these experiments) for smaller problems. On the other hand, the time required to find the optimum solution exhibits a very high variability, while SA is an anytime method that can usually find "good" solutions in a much more predictable amount of time. Experiments were run using dual-processor 850 MHz Linux boxes, and the Sun Java HotSpot compiler in client mode. Timings are given in wall-clock time.

The MAGNET system is written in Java, and the IP solver is lp_solve, available from ftp://ftp.ics.ele.tue.nl/pub/lp_solve/. Because we are using an out-of-process IP solver, some well-known techniques, such as starting with a subset of the constraint set and adding additional constraints only if they are violated, or recording multiple solutions as the solver runs, are not possible in our current environment.

Each problem set consists of 200 problems, with randomly-generated plans and randomly-generated bids. Our problem generator has a large number of parameters; we kept all of them constant except for the task-count, bid-count, and bid-size values. Since our goal is to evaluate bids in a time-limited multi-agent interaction situation, we are primarily interested in scalability and predictability. Secondarily, we are interested in discovering measurable problem characteristics that the agent can use to tune its evaluation process "on the fly."

The process for generating problems operates as follows:

1. *Generate plan*: The desired number of tasks is generated, and random precedence relations are created between them, avoiding redundant precedence links. Tasks are randomly selected from among three "task types" that specify different values of expected duration, duration variability, and expected resource availability.
2. *Compose RFQ*: An RFQ is generated by setting time windows for the tasks in the plan, and specifying the timeline for the bidding process. Time windows are set by determining the makespan of the plan (the longest path through the precedence network) and multiplying it by a "slack" factor of 1.2, then "relaxing" the time windows for individual tasks to allow some overlap. The final result is that individual tasks are given time windows of at least 125% of their expected values (tasks not on the critical path will have longer time windows).
3. *Generate Bids*: A specified number of attempts are made to generate bids against the RFQ. Each bid is generated by selecting a task at random from the plan, using the task-type parameters to generate a supplier time window for that task, and testing this time window against the time window specified in the RFQ for that task. If the supplier's time window is contained within the RFQ time window, we call it a "valid task spec" and add the task to the bid. If a valid task spec was generated, then with some probability, each predecessor and successor link from that task is followed to attempt to add additional tasks to the bid, and so on recursively. The resulting bids specify "contiguous" sets of tasks, and are guaranteed to be internally feasible. Finally, a cost is determined for the overall bid. Because valid task specs are not always achieved, some attempts to generate bids will fail altogether.

The first experiment compares the performance of the original IP formulation to the revised formulation. We were only able to run the original formulation on the smallest problems (5 tasks, up to 20 bids) because some problems were generating more than 10^5 rows and taking inordinate amounts of time to solve (we stopped one after 13 hours). Table 2 shows the results of this experiment. All entries are averaged across 200 problems. Key features to observe are the relative problem sizes (number of rows) and the extreme variability (given as σ(time)) of the original formulation. For the revised formulation, the reported time is the sum of preprocessing time and IP solver time. The time required for preprocessing is generally

TABLE 2
Comparing IP formulations.

		Original IP			Revised IP		
Task Count	Bid Count	Rows	Time (msec)	σ(time)	Rows	Time (msec)	σ(time)
5	11.4	487	195	1001	16.3	71.3	144
5	13.5	869	233	937	18.6	71.0	150
5	15.0	1265	753	5125	20.3	68.8	125
5	17.0	2271	3150	16256	22.7	95.8	279

between 2 and 20 times the run time of the IP solver itself, but this appears to be time well spent. Our experience attempting to solve larger problems with the original formulation shows that the advantage of preprocessing grows dramatically as problem size increases.

The second series demonstrates scalability of the search process, using both the revised IP formulation and our Simulated Annealing solver, as the size of the plan varies, with a (nearly) constant ratio of bids to tasks (the ratio varies somewhat due to the random nature of the bid-generation process). Tables 3 and 4 show problem characteristics for this set. In both tables, the "Solved" column gives the number of problems solved out of 200 (not all 200 problems were solvable). In Table 3 "Opt" shows the number of times the optimum solution was found, "Time" is the time taken and σ(time) gives the standard deviation of the "Time" column. It should be noted that the performance results for the SA search are somewhat arbitrary; the annealing schedule and stopping conditions can be adjusted at will with respect to any measurable characteristic of the problem. Clearly, the SA search would have achieved better optimization results on the larger problems if we scaled it more strongly with respect to problem size. On the other hand, the SA search does often find optimal solutions, even though there is no way to know they are optimal. The non-optimal solutions it reports are typically within a few percent of optimal, even though the typical feasible solution in this environment is about 50% worse than optimal.

In Table 4, the times for preprocessing and for the IP solver are given separately, as "PP" and "IP". The σ(time) column gives the standard deviation of the sum of these two times. The individual variabilities of the PP time and the IP time are comparable – the large values of σ in the table are not primarily due to either process.

Figure 2 shows graphically the relative average performance of the SA and IP approaches on the task-size experiment. The striking feature of this graph is the apparent improvement in performance of the SA approach as problem size increases. This is an artifact of the tuning parameters, which

TABLE 3
Task size experiment: Simulated Annealing.

Task Count	Bid Count	Bid Size	Solved	Opt	Time (msec)	σ(time)
5	13.5	2.14	188	188	10	10
10	28.5	3.20	184	178	147	391
15	45.4	4.35	179	141	524	986
20	60.8	5.21	169	118	1171	1707
25	77.6	6.30	147	89	1958	2683
30	93.3	7.41	141	80	3445	4012
35	110.6	8.69	116	66	4012	4289

TABLE 4
Task size experiment: Revised IP.

Task Count	Bid Count	Bid Size	Solved	Rows	PP (msec)	IP (msec)	σ(time)
5	13.5	2.14	188	19.1	10.4	34.5	15
10	28.5	3.20	184	44.3	70.1	34.3	57
15	45.4	4.35	181	85.6	191.0	30.2	170
20	60.8	5.21	176	145.0	489.0	40.7	380
25	77.6	6.30	165	281.0	1268.0	56.2	1423
30	93.3	7.41	157	514.0	3375.0	194.0	4353
35	110.6	8.69	137	861.0	5181.0	317.0	7100

are cutting off the search too soon on the larger problems. We can clearly see the corresponding dropoff in optimization performance in Table 3.

In Figure 3, we show the result of varying the bid count with a fixed plan size of 20 tasks. Here, the exponential tendency of the IP method is clear, while the time required by the SA method is close to linear. This is accompanied by a corresponding fall-off of optimization performance of the SA method as the density of solutions increases. We see this in the data set labeled "SA opt", which plots the ratio of the optimum solution as determined by the IP search, to the solution reported by the SA method.

Finally, in Figure 4, we see the effect on search effort as the average size of bids is varied. For this set, the task count was held constant at 20 tasks, and the bid count was 70 bids. As in the previous set, the optimization performance of the SA method is traded off against run time to produce these results.

It is apparent from these experiments and others that the probability of finding the optimal solution in a given amount of time is higher with the

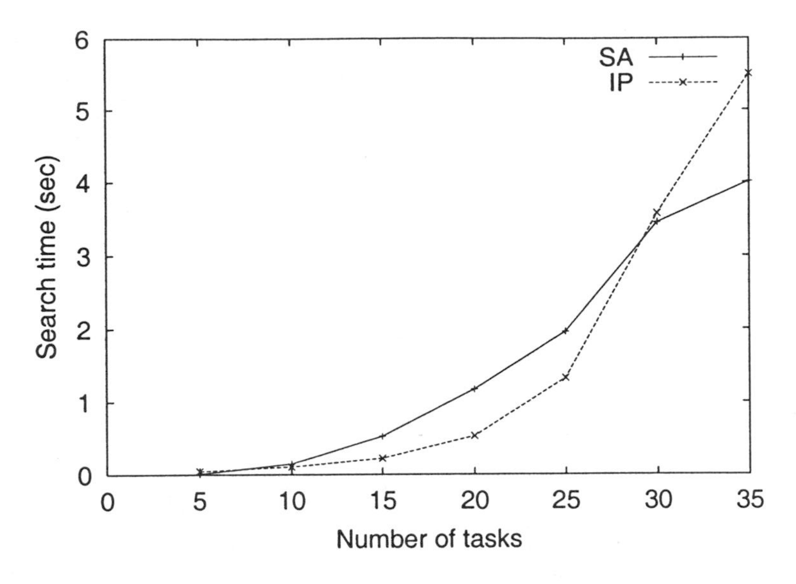

FIG. 2. *Search time as a function of task count.*

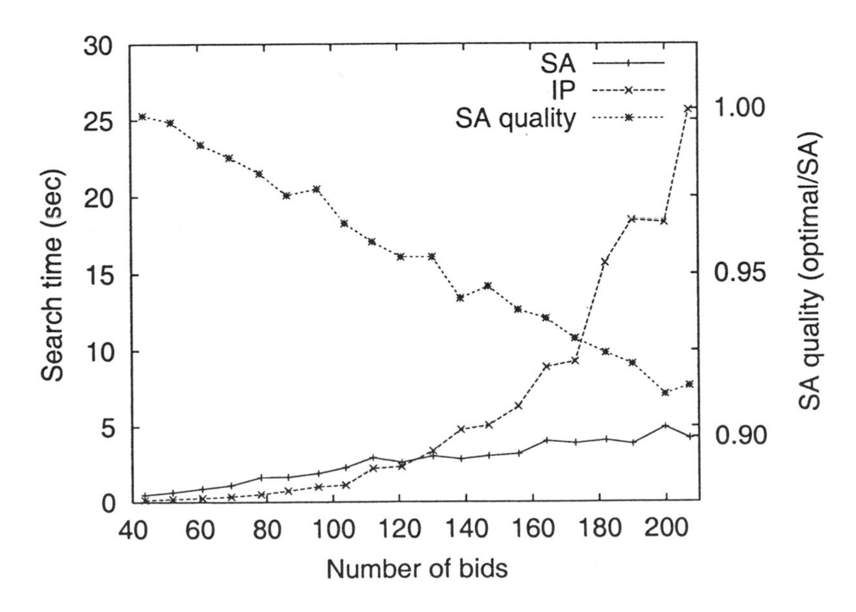

FIG. 3. *Search time as a function of bid count.*

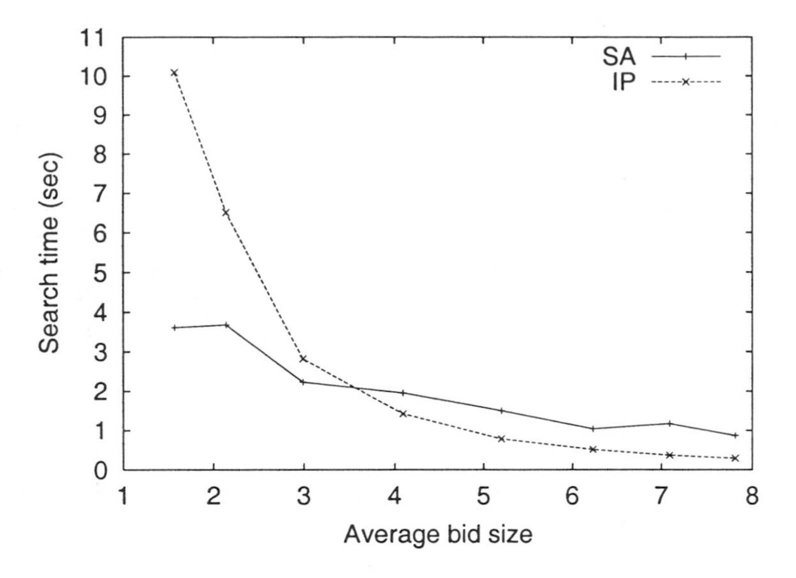

FIG. 4. *Search time as a function of bid size.*

IP approach than with the SA approach. For the purpose of supporting an agent involved in a negotiation process, the key difference between the SA and IP search methods seems to be in controllability and predictability. Both methods exhibit a high variability in the amount of time required to produce a solution, and no significant correlations between overall problem characteristics and the amount of time required have been found. Nor is there a strong correlation between the quality of the SA result and the time required for the IP solution. More significantly, perhaps, there is no correlation (correlation coefficient < 0.1 on all problem sets) between the IP solution time and the SA solution time even for problems in which SA found the optimal solution. For an agent that must make a decision in a fixed amount of time, the extreme variability of either approach presents an unacceptable situation. It appears that an ideal approach would be to run both methods in parallel. Both the IP method and the SA method can be configured to deliver candidate solutions "on the fly", and SA solutions can be used to bound the IP search. This can be especially significant in the cases where preprocessing time dominates the IP solution time.

The results given here for the Simulated Annealing search method do not compare well with the results reported in [4], because we have eliminated the notion of bid break-downs. In the original system, there was an assumption that suppliers who submitted bids on multiple tasks were also obligated to perform any subset of those tasks. This allowed the search engine to "take apart" the bids it was given, and greatly simplified the problem of finding a feasible, if suboptimal solution. Without the bid breakdown assumption, the problem has become significantly more difficult to solve, and the version of the Simulated Annealing search reported here is tuned to anneal more slowly and to search as much as 10 times longer before terminating.

5. Related work. The determination of winners of combinatorial auctions [8] is known to be hard. Many algorithms have been proposed to produce good or optimal allocations. Dynamic programming [14] produces optimal allocations, but works well only for small sets of bids, and imposes significant restrictions on the class of bids. Nisan [10] formalizes several bidding languages and compares their expressive power. He analyzes different classes of auctions, and proposes an approach based on Linear Programming for bid allocation. Shoham [5] produces optimal allocations for OR-bids with dummy items by cleverly pruning the search space. Sandholm [15, 17] uses an anytime algorithm to produce optimal allocations for a more general class of bids, which includes XOR and OR-XOR bids. Andersson [1] proposes integer programming for winner determination in combinatorial auctions. The major difference is that in the cases studied for combinatorial auctions, bid allocation is determined solely by cost. Our setting is more general. Our agents have to cover all the tasks, ensure feasibility of the bids they accept, and reduce scheduling risk.

Walsh has proposed using combinatorial auction mechanisms for supply-chain formation [18] and for decentralized scheduling [19]. Neither of these proposals requires the allocation solver to deal with temporal feasibility, which is the principal problem dealt with in this work.

One of the algorithms we used is based on simulated annealing [13], and as such combines the advantages of heuristically guided search with some random search. Since the introduction of iterative sampling [7], a strategy that randomly explores different paths in a search tree, there have been numerous attempts to improve search performance by using randomization. A variety of methods that combine randomization with heuristics have been proposed, such as Least Discrepancy Search [6], heuristic-biased stochastic sampling [2], and stochastic procedures for generating feasible schedules [11], just to name a few.

6. Conclusion and future work. This work was motivated by the need for an improved bid evaluation procedure for the MAGNET system, and by the recent successes reported in applying Integer Programming to the related Combinatorial Auction winner determination problem. We have shown that Integer Programming can be applied to the problem of bid evaluation in a combinatorial auction situation that includes temporal constraints among items, and lacks the free-disposal option. The formulation we present typically requires an amount of pre-processing that is significantly greater than the time required by the IP solver itself. A significant weakness of the IP approach is the high degree of variability in run time.

For an agent that must perform on a fixed time schedule, this may not be acceptable. We suggest that running a stochastic search in parallel with an IP solver may alleviate this drawback, and provide experimental data that supports this approach.

Several open questions remain. Many IP solution methods require a much tighter integration between the host application and the IP solver than we have been able to implement. This would allow the application to set bounds, add constraints, and record interim (potentially suboptimal) integer solutions. Such an integration would allow exploration of the potential synergy between a heuristic search engine and an IP solver in time-constrained situations such as a MAGNET agent must face. In addition, the data reported here make it clear that the Simulated Annealing search method needs to be better-tuned across the range of problem sizes. Both the stopping criteria and the annealing schedule need to be better adjusted in response to multiple problem-size measures. We have explored several problem-size measures in this work, but others might also be important, such as the degree of variability in bid size, task coverage, or cost/task.

It seems clear that further simplifications could be achieved in preprocessing. For example, when a singleton is encountered, it could tighten the time constraints, and the effects could be propagated through the task network. We would also like to understand how to incorporate risk factors that depend on the distribution of slack in the schedule, and how to optimize completion time without significantly increasing the complexity of the problem.

REFERENCES

[1] A. Andersson, M. Tenhunen, and F. Ygge, *Integer programming for combinatorial auction winner determination*, in Proc. of 4th Int'l Conf on Multi-Agent Systems, IEEE Computer Society Press, July 2000, pp. 39–46.

[2] J.L. Bresina, *Heuristic-biased stochastic sampling*, in Proc. of the Thirteenth Nat'l Conf. on Artificial Intelligence, 1996.

[3] J. Collins, C. Bilot, M. Gini, and B. Mobasher, *Mixed-initiative decision support in agent-based automated contracting*, in Proc. of the Fourth Int'l Conf. on Autonomous Agents, June 2000, pp. 247–254.

[4] J. Collins, R. Sundareswara, M. Gini, and B. Mobasher, *Bid selection strategies for multi-agent contracting in the presence of scheduling constraints*, in Agent Mediated Electronic Commerce II, A. Moukas, C. Sierra, and F. Ygge, eds., vol. LNAI1788, Springer-Verlag, 2000.

[5] Y. Fujishjima, K. Leyton-Brown, and Y. Shoham, *Taming the computational complexity of combinatorial auctions: Optimal and approximate approaches*, in Proc. of the 16th Joint Conf. on Artificial Intelligence, 1999.

[6] W.D. Harvey and M.L. Ginsberg, *Limited discrepancy search*, in Proc. of the 14th Joint Conf. on Artificial Intelligence, 1995, pp. 607–613.

[7] P. Langley, *Systematic and nonsystematic search strategies*, in Proc. Int'l Conf. on AI Planning Systems, College Park, Md, 1992, pp. 145–152.

[8] R. McAfee and P.J. McMillan, *Auctions and bidding*, Journal of Economic Literature, 25 (1987), pp. 699–738.

[9] P. MILGROM, *Auction and bidding: a primer*, Journal of economic perspectives, **3** (1989), pp. 3–22.
[10] N. NISAN, *Bidding and allocation in combinatorial auctions*, in Proc. of ACM Conf on Electronic Commerce (EC'00), Minneapolis, Minnesota, October 2000, ACM SIGecom, ACM Press, pp. 1–12.
[11] A. ODDI AND S.F. SMITH, *Stochastic procedures for generating feasible schedules*, in Proc. of the Fourteenth Nat'l Conf. on Artificial Intelligence, 1997, pp. 308–314.
[12] M.E. POLLACK, *Planning in dynamic environments: The DIPART system*, in Advanced Planning Technology, A. Tate, ed., AAAI Press, 1996.
[13] C.R. REEVES, *Modern Heuristic Techniques for Combinatorial Problems*, John Wiley & Sons, New York, NY, 1993.
[14] M.H. ROTHKOPF, A. PEKEČ, AND R.M. HARSTAD, *Computationally manageable combinatorial auctions*, Management Science, **44** (1998), pp. 1131–1147.
[15] T. SANDHOLM, *An algorithm for winner determination in combinatorial auctions*, in Proc. of the 16th Joint Conf. on Artificial Intelligence, 1999, pp. 524–547.
[16] ———, *Approaches to winner determination in combinatorial auctions*, Decision Support Systems, **28** (2000), pp. 165–176.
[17] T. SANDHOLM AND S. SURI, *Improved algorithms for optimal winner determination in combinatorial auctions and generalizations*, in Proc. of the Seventeen Nat'l Conf. on Artificial Intelligence, 2000, pp. 90–97.
[18] W.E. WALSH, M. WELLMAN, AND F. YGGE, *Combinatorial auctions for supply chain formation*, in Proc. of ACM Conf on Electronic Commerce (EC'00), October 2000.
[19] W.E. WALSH, M.P. WELLMAN, P.R. WURMAN, AND J.K. MACKIE-MASON, *Some economics of market-based distributed scheduling*, in Proc. of the Eighteenth Int'l Conf. on Distributed Computing Systems, 1998, pp. 612–621.

LINEAR PROGRAMMING AND VICKREY AUCTIONS*

SUSHIL BIKHCHANDANI†, SVEN DE VRIES‡,
JAMES SCHUMMER§, AND RAKESH V. VOHRA¶

Abstract. The Vickrey sealed bid auction occupies a central place in auction theory because of its efficiency and incentive properties. Implementing the auction requires the auctioneer to solve $n+1$ optimization problems, where n is the number of bidders. In this paper we survey various environments (some old and some new) where the payments bidders make under the Vickrey auction correspond to dual variables in certain linear programs. Thus, in these environments, at most two optimization problems must be solved to determine the Vickrey outcome. Furthermore, primal-dual algorithms for some of these linear programs suggest ascending auctions that implement the Vickrey outcome.

Key words. Vickrey auctions, multi-item auctions, combinatorial auctions, duality, primal-dual algorithm.

AMS(MOS) subject classifications. 91B24, 91B26, 90C27.

1. Introduction.

The Vickrey sealed bid auction deservedly occupies an important place in the literature on mechanism design. It results in an economically efficient outcome, and it is a (weakly) dominant strategy for bidders to bid truthfully.

To describe the auction, let N be the set of n agents (or bidders), and F be the set of feasible allocations among them. Each agent $j \in N$ assigns a monetary benefit $v_j(a)$ to each allocation $a \in F$. This monetary benefit v_j is private information to agent j. The efficient outcome is the outcome a^* that solves $\max_{a\in F}\sum_{j\in N} v_j(a)$. Let

$$V(N) = \max_{a\in F}\sum_{j\in N} v_j(a) \text{ and}$$
$$V(N\setminus j) = \max_{a\in F}\sum_{i\in N\setminus j} v_i(a).$$

The sealed bid auction proceeds as follows. Agents submit their benefit functions $\{v_1, v_2, \ldots, v_n\}$. For the moment, assume they do so truthfully; subsequently, we show that doing otherwise cannot improve any agent's payoff. The auctioneer computes a^*, $V(N)$ and $V(N\setminus j)$ for all $j \in N$.

*We gratefully acknowledge invaluable discussions with Joe Ostroy at U.C.L.A.

†Anderson Graduate School of Management, UCLA, Los Angeles, CA 90095.

‡Zentrum Mathematik, TU München, D-80290 München, Germany.

§Department of Managerial Economics and Decision Sciences, Kellogg Graduate School of Management, Northwestern University, Evanston IL 60208.

¶Department of Managerial Economics and Decision Sciences, Kellogg Graduate School of Management, Northwestern University, Evanston IL 60208.

The allocation a^* is chosen, and bidder j pays

$$V(N \setminus j) - \sum_{i \neq j} v_i(a^*).$$

Thus, bidder j's net benefit from participating is

$$v_j(a^*) - \left[V(N \setminus j) - \sum_{i \neq j} v_i(a^*)\right] = V(N) - V(N \setminus j).$$

This last term is sometimes called bidder j's *marginal product*, and is clearly non-negative. It is the incremental benefit from including agent j in the auction.

Could any bidder benefit from misreporting his valuation function? No. To see why, suppose bidder j reports $u_j \neq v_j$ instead. The auctioneer then chooses an $a' \in F$ that solves $\max_{a \in F}[\sum_{i \neq j} v_i(a) + u_j(a)]$. Agent j's payoff becomes

$$\left[\sum_{i \neq j} v_i(a') + v_j(a')\right] - V(N \setminus j) \leq V(N) - V(N \setminus j).$$

A closer inspection of the argument reveals more: No matter what the other agents report, no agent can benefit by misreporting his valuation function.

The Vickrey auction is not the only efficient auction with the property that it is a (weakly) dominant strategy for bidders to bid truthfully. The general class of such schemes is the class of Clark–Groves mechanisms.[1] The Vickrey auction is one member in this class and, in some sense, is focal. For instance, Krishna and Perry (1998) justify its use in terms of revenue maximization within a class of mechanisms.[2]

Computationally, the implementation of a Vickrey auction requires the solution of $n+1$ optimization problems. In many environments of economic interest, each of these problems is a linear program. An agent makes its presence known in the choice of constraints (rows) or variables (columns) of the linear program. It is therefore tempting to think that the marginal product of an agent should be encoded in the optimal dual variables of the linear program. After all, it is these variables that inform us of the effect of changing the right hand side of a constraint, or modifying a coefficient in the objective function.

In this paper, we survey the instances where such a connection exists. We start in Section 2 with a discussion of the general problem, and various formulations of it, building on the notions of Bikhchandani and Ostroy (2000a). In the sections that follow, we examine various special cases of the general model which admit simpler formulations.

[1] See Vickrey (1961), Clarke (1971), and Groves (1973).
[2] See also, Williams (1999). Moulin (1986) justifies its use in a different sense.

The Vickrey auction as we have just described is a sealed bid auction. In the auction of a single object, it has long been known that an open (ascending) auction exists that duplicates the outcomes of the Vickrey auction. This is the English auction, in which the auctioneer continuously raises the price for the object. An agent is to keep his hand raised until the price exceeds the amount the agent is willing to pay for the object. The auction terminates as soon as only one agent's hand is left raised. That agent wins the object at that terminal price.

In such an auction, as long as every other agent is behaving in a way consistent with some valuation for the object, an agent cannot benefit by lowering his hand prematurely, or by keeping his hand raised beyond the price he is willing to pay. So, the ascending auction has desirable incentive properties comparable to its sealed bid counterpart, and a number of other pleasant features that are listed by Ausubel (1997).

Here, we point out that such ascending auctions can be interpreted as particular implementations of a primal dual algorithm for the underlying optimization problem. Specific examples of this are provided by Ausubel (1997) and by Demange, Gale, and Sotomayor (1986). Such a connection should not be surprising; an ascending auction that implements the Vickrey outcome must solve the same optimization problem as an efficient algorithm, and the auction *is* an algorithm.

2. The general case. We begin with the general case of an auctioneer who must auction off a set of distinct heterogeneous objects, M. Let N be the set of bidders. For every set of objects $S \subseteq M$, let $v_j(S)$ be the value that agent $j \in N$ assigns to consuming S.[3] An *allocation* is an assignment of objects to the agents. To formulate the problem of finding an *efficient* allocation let $y(S, j) = 1$ if the bundle $S \subseteq M$ is allocated to $j \in N$ and zero otherwise.[4] The optimization problem, denoted **(P1)**, is to solve

$$V(N) = \max \sum_{j \in N} \sum_{S \subseteq M} v_j(S) y(S, j)$$

$$\text{s.t.} \sum_{S \ni i} \sum_{j \in N} y(S, j) \leq 1 \quad \forall i \in M$$

$$\sum_{S \subseteq M} y(S, j) \leq 1 \quad \forall j \in N$$

$$y(S, j) = 0, 1 \quad \forall S \subseteq M, \forall j \in N.$$

The first constraint ensures that overlapping sets of goods are never assigned. The second ensures that no bidder receives more than one subset.

---

[3] A more general specification is to define agents' valuations over allocations, allowing for externalities: preferences affected by *others'* consumption. For simplicity, we do not specify such a general model.

[4] It is important to make clear the notation that, e.g., if agent j consumes the pair of distinct objects $i, i' \in M$, then $y(\{i, i'\}, j) = 1$, but $y(\{i\}, j) = 0$. An agent consumes exactly one set of objects.

The linear relaxation of the above integer program admits fractional solutions, so it is natural to look for a stronger formulation. The way we do this is by using auxiliary variables. It is helpful to first consider the formulation (P2) below. This stronger formulation is not strong enough, but it suggests directions for other formulations; hence we introduce it.

Let Π be the set of all possible partitions of the objects in the set M. For any partition $\sigma \in \Pi$, we write $S \in \sigma$ to mean that the set $S \subset M$ is a part of the partition σ. Auxiliary variables are denoted z_σ, where $z_\sigma = 1$ if the partition σ is selected and $z_\sigma = 0$ otherwise. Using them we obtain formulation **(P2)**.

$$
\begin{aligned}
V(N) = \max & \sum_{j\in N}\sum_{S\subseteq M} v_j(S)y(S,j) \\
\text{s.t. } & \sum_{S\subseteq M} y(S,j) \le 1 \quad \forall j \in N \\
& \sum_{j\in N} y(S,j) \le \sum_{\sigma\ni S} z_\sigma \quad \forall S \subset M \\
& \sum_{\sigma\in\Pi} z_\sigma \le 1 \\
& y(S,j) = 0,1 \quad \forall S \subseteq M, \forall j \in N.
\end{aligned}
$$

This formulation chooses a partition of M and then assigns the sets of the partition to bidders in such a way as to maximize total valuations. It is easy to see that formulation (P2) is stronger than (P1): Fix an $i \in M$ and add over all $S \ni i$ the inequalities

$$\sum_{j\in N} y(S,j) \le \sum_{\sigma\ni S} z_\sigma \quad \forall S \subset M$$

to obtain

$$\sum_{S\ni i}\sum_{j\in N} y(S,j) \le 1 \quad \forall i \in M$$

which are the inequalities that appear in the first formulation.

Even (P2) admits fractional solutions in its linear relaxation. The easiest way to see this is to consider the following.

$$
\begin{aligned}
F(z) = \max & \sum_{j\in N}\sum_{S\subseteq M} v_j(S)y(S,j) \\
\text{s.t. } & \sum_{S\subseteq M} y(S,j) \le 1 \quad \forall j \in N \\
& \sum_{j\in N} y(S,j) \le \sum_{\sigma\ni S} z_\sigma \quad \forall S \subset M \\
& y(S,j) \ge 0 \quad \forall S \subseteq M, \forall j \in N.
\end{aligned}
$$

$F(z)$ is a concave function of z. Hence

$$V(N) = \max\{F(z) : \text{s.t. } \sum_{\sigma \in \Pi} z_\sigma \leq 1\}.$$

Clearly, the optimal choice of z need not occur at an extreme point.

Bikhchandani and Ostroy (2000a) propose an even stronger formulation that is integral. Furthermore they identify a necessary and sufficient condition for the dual variables to correspond to the Vickrey payoffs. This formulation we consider next.

2.1. The Bikhchandani and Ostroy formulation. Bikhchandani and Ostroy (2000a) introduce a variable for every feasible integer solution. Let μ denote both a partition of the set of objects *and* an assignment of the elements of the partition to bidders. Thus μ and μ' can give rise to the same partition, but to different assignments of the parts to bidders. Let Γ denote the set of all such partition-assignment pairs. We will write $S^j \in \mu$ to mean that under μ, agent j receives the set S. Let $\delta_\mu = 1$ if the partition-assignment pair $\mu \in \Gamma$ is selected, and zero otherwise. Using these new variables the efficient allocation can be found by solving

$$\begin{aligned} V(N) = \max & \sum_{\mu \in \Gamma} \left[\sum_{j \in N} \sum_{S^j \in \mu} v_j(S^j) \right] \delta_\mu \\ \text{s.t. } & \sum_{\mu \in \Gamma} \delta_\mu \leq 1 \\ & \delta_\mu \geq 0 \quad \forall \mu \in \Gamma. \end{aligned}$$

This formulation involves one constraint and so gives rise to one dual variable only. Hence it is not detailed enough to provide information about the marginal products of each of the bidders. What is needed is a disaggregate formulation that allows one to track the effect of removing or adding a bidder. Here is the one proposed by Bikhchandani and Ostroy (2000a), which we call **(P3)**.

$$\begin{aligned} V(N) = \max & \sum_{j \in N} \sum_{S \subseteq M} v_j(S) y(S,j) \\ \text{s.t. } & y(S,j) \leq \sum_{\mu \ni S^j} \delta_\mu \quad \forall j \in N, \forall S \subseteq M \\ & \sum_{S \subseteq M} y(S,j) \leq 1 \quad \forall j \in N \\ & \sum_{\mu \in \Gamma} \delta_\mu \leq 1 \\ & y(S,j) = 0,1 \quad \forall S \subseteq M, \forall j \in N. \end{aligned}$$

A straightforward argument shows that the formulation has the integrality property.

The significance of this formulation is that the dual variable associated with the second constraint, $\sum_{S \subseteq M} y(S, j) \leq 1$, can be interpreted as agent j's marginal product. To see why, for an agent $j \in N$, reduce the right hand side of the corresponding constraint to zero. This has the effect of removing agent j from the problem. The resulting change in optimal objective function value will be agent j's marginal product.

The difficulty is to ensure that a dual solution exists such that, simultaneously, each dual variable associated with that type of constraint takes on the value of the corresponding agent's marginal product. Bikhchandani and Ostroy (2000a) derive the following necessary and sufficient condition for this to be realized.

$$V(N) - V(K) \geq \sum_{j \in N \setminus K} [V(N) - V(N \setminus j)] \quad \forall K \subseteq N. \tag{1}$$

They interpret this condition to mean "agents are substitutes."

This condition formalizes the notion that the contribution (i.e., marginal product) of a group of agents is more than the sum of the contributions of the individual members of the group. Such a condition motivates, for example, the idea that workers are better-off forming a union rather than bargaining separately with management.

To write down the dual, we associate with each constraint $y(S, j) \leq \sum_{\mu \ni S^j} \delta_\mu$ a variable $p_j(S) \geq 0$ which can be interpreted as the price that agent j pays for the set S. To each constraint $\sum_{S \subseteq M} y(S, j) \leq 1$ we associate a variable $\pi_j \geq 0$ which can be interpreted as agent j's surplus. To the constraint $\sum_{\mu \in \Gamma} \delta_\mu \leq 1$ we associate the variable π^s which can be interpreted as the seller's surplus. The dual **(DP3)** becomes

$$\begin{aligned}
&\min \sum_{j \in N} \pi_j + \pi^s \\
&\text{s.t. } p_j(S) + \pi_j \geq v_j(S) \quad \forall j \in N, \forall S \subseteq M \\
&\qquad - \sum_{S^j \in \mu} p_j(S) + \pi^s \geq 0 \quad \forall \mu \in \Gamma \\
&\qquad p_j(S) \geq 0 \quad \forall j \in N, \forall S \subseteq M
\end{aligned}$$

THEOREM 1. *Suppose the "agents are substitutes" condition (1) holds. Then there exists an optimal solution to (DP3), $((\pi_j), \pi^s)$, where $\pi_j = V(N) - V(N \setminus j)$ for all $j \in N$. That is, agent j's surplus, π_j, is agent j's marginal product. The converse is also true.*

Proof. Set $\pi_j = V(N) - V(N \setminus j)$ for all $j \in N$. By setting

$$\begin{aligned}
p_j(S) &= [v_j(S) - \pi_j]^+ \quad \forall S \subseteq M, \forall j \in N \\
\pi^s &= \max_{\mu \in \Gamma} \sum_{S^j \in \mu} p_j(S)
\end{aligned}$$

we obtain a feasible solution to the dual. Let μ^* be the partition-assignment pair that defines π^s above. The objective function value of our feasible dual solution is

$$\sum_{j \in N} \pi_j + \sum_{S^j \in \mu^*} [v_j(S^j) - \pi_j]^+.$$

To obtain a contradiction suppose that

$$\sum_{j \in N} \pi_j + \sum_{S^j \in \mu^*} [v_j(S^j) - \pi_j]^+ > V(N).$$

Let $K = \{j : [v_j(S^j) - \pi_j]^+ > 0\}$. Then the previous inequality becomes

$$\sum_{j \in N} \pi_j - \sum_{j \in K} \pi_j + \sum_{j \in K} v_j(S^j) > V(N).$$

Since $\sum_{j \in K} v_j(S^j) \leq V(K)$, the last inequality implies

$$V(N) - V(K) < \sum_{j \in N \setminus K} [V(N) - V(N \setminus j)],$$

which is a violation of the substitutes condition. Running the argument in reverse yields the equivalence. □

2.2. A new formulation. Here we give a new formulation that uses fewer variables than the Bikhchandani and Ostroy formulation. It consists of a collection of assignment problems linked together by two classes of constraints.

As before, let Π be the set of partitions of M. For any partition $\sigma \in \Pi$, $S \in \sigma$ means that $S \subset M$ is a part of the partition σ. Let $z_\sigma = 1$ if the partition σ is selected and zero otherwise. Set $y^\sigma(S, j) = 1$ to mean that partition σ is selected and $S \in \sigma$ is assigned to agent j, and set it equal to zero otherwise. We call this formulation **(P4)**.

$$V(N) = \max \sum_{\sigma \in \Pi} \sum_{S \in \sigma} \sum_{j \in N} v_j(S) y^\sigma(S, j)$$

$$\text{s.t.} \sum_{S \in \sigma} y^\sigma(S, j) \leq z_\sigma \quad \forall j \in N, \forall \sigma \in \Pi$$

$$\sum_{j \in N} y^\sigma(S, j) \leq z_\sigma \quad \forall S \in \sigma, \forall \sigma \in \Pi$$

$$\sum_{\sigma \in \Pi} z_\sigma \leq 1$$

$$\sum_{\sigma \in \Pi} \sum_{S \in \sigma} y^\sigma(S, j) \leq 1 \quad \forall j \in N$$

$$y^\sigma(S, j) \geq 0 \quad \forall S \in \sigma, \forall j \in N$$

$$z_\sigma \geq 0 \quad \forall \sigma \in \Pi.$$

The constraints $\sum_{\sigma\in\Pi}\sum_{S\in\sigma} y^{\sigma}(S,j) \leq 1$ are redundant. Their dual, however, corresponds to agent j's marginal product.

THEOREM 2. *Formulation (P4) has an optimal integral solution.*

Proof. In what follows we ignore the redundant constraint. Let

$$
\begin{aligned}
G(z_\sigma) = \max & \sum_{S\in\sigma}\sum_{j\in N} v_j(S) y^{\sigma}(S,j) \\
\text{s.t. } & \sum_{S\in\sigma} y^{\sigma}(S,j) \leq z_\sigma \quad \forall j \in N \\
& \sum_{j\in N} y^{\sigma}(S,j) \leq z_\sigma \quad \forall S \in \sigma \\
& y^{\sigma}(S,j) \geq 0 \quad \forall S \in \sigma, \forall j \in N.
\end{aligned}
$$

Notice that each $G(z_\sigma)$ is the optimal objective function value of an assignment problem with all right hand sides being some number z_σ. Then

$$
\begin{aligned}
V(N) = \max & \sum_{\sigma\in\Pi} G(z_\sigma) \\
\text{s.t. } & \sum_{\sigma\in\Pi} z_\sigma \leq 1 \\
& 0 \leq z_\sigma \leq 1 \quad \forall \sigma \in \Pi
\end{aligned}
$$

Observe that $G(z_\sigma)$ is linear in z_σ. Thus the optimal solution will have z_σ being integral. □

We now formulate the dual to (P4) and call it **(DP4)**. Let $u_j^{\sigma} \geq 0$ be the dual variables associated with the constraints

$$\sum_{S\in\sigma} y^{\sigma}(S,j) \leq z_\sigma \quad \forall j \in N.$$

Let $w_S^{\sigma} \geq 0$ be the dual variables associated with the constraints:

$$\sum_{j\in N} y^{\sigma}(S,j) \leq z_\sigma \quad \forall S \in \sigma.$$

Let $\pi^s \geq 0$ be the dual variable associated with the constraint

$$\sum_{\sigma\in\Pi} z_\sigma \leq 1.$$

Let $\pi_j \geq 0$ be the dual variable associated with the constraint

$$\sum_{\sigma\in\Pi}\sum_{S\in\sigma} y^{\sigma}(S,j) \leq 1 \quad \forall j \in N.$$

With these variables, (DP4) is

$$V(N) = \min \sum_{j \in N} \pi_j + \pi^s$$
$$\text{s.t. } \pi^s \geq \sum_{S \in \sigma} w_S^\sigma + \sum_{j \in N} u_j^\sigma \quad \forall \sigma \in \Pi$$
$$\pi_j + u_j^\sigma + w_S^\sigma \geq v_j(S) \quad \forall j \in N, \forall S \in \sigma$$
$$\pi_j, u_j^\sigma, w_S^\sigma, \pi^s \geq 0 \quad \forall j \in N, \forall S \in \sigma, \forall \sigma \in \Pi$$

The analogous result to Theorem 1 is the following.

THEOREM 3. *Suppose the "agents are substitutes" condition (1) holds. Then there exists an optimal solution to (DP4), $((\pi_j), \pi^s)$, where $\pi_j = V(N) - V(N \setminus j)$ for all $j \in N$. The converse is also true.*

Proof. Set $\pi_j = V(N) - V(N \setminus j)$ for all $j \in N$, and for each $\sigma \in \Pi$, choose u^σ and w^σ so as to solve:

$$\min \sum_{j \in N} u_j^\sigma + \sum_{S \in \sigma} w_S^\sigma$$
$$\text{s.t. } u_j^\sigma + w_S^\sigma \geq [v_j(S) - \pi_j]^+ = v_j'(S)$$

This is just the dual to the following assignment problem:

$$H(\sigma) = \max \sum_{j \in N} \sum_{S \in \sigma} v_j'(S) y(S, j)$$
$$\text{s.t. } \sum_{S \in \sigma} y^\sigma(S, j) \leq 1 \quad \forall j \in N$$
$$\sum_{j \in N} y^\sigma(S, j) \leq 1 \quad \forall S \in \sigma$$
$$0 \leq y^\sigma(S, j) \quad \forall S \in \sigma, \forall j \in N$$

Set $\pi^s = \max_{\sigma \in \Pi} H(\sigma)$. Let the maximum be attained on the partition σ^*. Then the objective function value of (DP4) is

$$\sum_{j \in N} \pi_j + H(\sigma^*).$$

For a contradiction, assume that

$$\sum_{j \in N} \pi_j + H(\sigma^*) > V(N).$$

In the partition σ^* let K be the set of agents who receive an element of σ^* as specified in $H(\sigma^*)$. Let S^j be the element assigned to agent j. Then

$$\begin{aligned} V(N) &< \sum_{j \in N} \pi_j + \sum_{j \in K} v_j'(S^j) = \sum_{j \in N} \pi_j + \sum_{j \in K} v_j(S^j) - \sum_{j \in K} \pi_j \\ &\leq V(K) + \sum_{j \in N \setminus K} \pi_j, \end{aligned}$$

which violates the substitutes condition.

To prove the converse, if the partition-assignment pair μ contains the partition σ, we write $\mu \in \sigma$. Let $p_j^\sigma(S) \equiv u_j^\sigma + w_S^\sigma$, and $P^\sigma \equiv \sum_{S^j \in \mu, \mu \in \sigma} p_j^\sigma(S)$. Thus, we may write (DP4) as:

$$
\begin{aligned}
V(N) = \min \sum_{j \in N} \pi_j + \pi^s \\
\text{s.t. } \pi^s - P^\sigma \geq 0 \ \forall \sigma \in \Pi \\
\pi_j + p_j^\sigma(S) \geq v_j(S) \quad \forall j \in N, \forall S \in \sigma, \forall \sigma \in \Pi \\
\pi_j, u_j^\sigma, w_S^\sigma \geq 0 \quad \forall j \in N, \forall S \in \sigma, \forall \sigma \in \Pi
\end{aligned}
$$

Let $(\pi_j, \pi^s, p_j^\sigma(S))$ be a feasible solution to (DP4), and define $p_j(S) \equiv \min_\sigma p_j^\sigma(S)$. Clearly, $(\pi_j, \pi^s, p_j(S))$ satisfies the buyer maximization constraints in (DP3); the satisfaction of the seller maximization constraints in (DP3) follows from the fact that for any $\mu \in \sigma$, $P^\sigma \geq \sum_{S^j \in \mu} p_j(S)$. Hence the feasible region of (DP4) is a subset of the feasible region of (DP3).

Thus by Theorem 1, if agents are not substitutes then there does not exist a (DP4) optimal solution which gives each buyer his marginal product. □

It is interesting to observe that the dual variables in (DP3) can be related to the ones in (DP4) by setting

$$p_j(S) = \min_{\sigma \in \Pi} (u_j^\sigma + w_S^\sigma).$$

This allows us to break the agents' payments into a non-anonymous component u_j^σ and a non-linear component w_S^σ.

2.3. On accelerated computation. In order to compute the marginal products of agents using (P4), we would have to solve two optimization problems. The first is (P4) itself to determine $V(N)$. The second is

$$
\begin{aligned}
\max \sum_{j \in N} \pi_j \\
\text{s.t. } V(N) = \sum_{j \in N} \pi_j + \pi^s \\
\pi^s \geq \sum_{S \in \sigma} w_S^\sigma + \sum_{j \in N} u_j^\sigma \quad \forall \sigma \in \Pi \\
\pi_j + u_j^\sigma + w_S^\sigma \geq v_j(S) \quad \forall j \in S, \forall S \in \sigma \\
\pi_j, u_j^\sigma, w_S^\sigma, \pi^s \geq 0 \quad \forall j \in N, \forall S \in \sigma, \forall \sigma \in \Pi
\end{aligned}
$$

In words, among all optimal solutions to (DP4) find the one that maximizes $\sum_{j \in N} \pi_j$.

With some prior knowledge about the range of valuations, one can reduce the computation to a single optimization problem. Standard arguments from linear programming show that there must be $\epsilon > 0$ sufficiently small such that the optimal solution to the previous linear program coincides with the following one.

$$\min \sum_{j \in N} \pi_j + (1+\epsilon)\pi^s$$
$$\text{s.t. } \pi^s \geq \sum_{S \in \sigma} w_S^\sigma + \sum_{j \in N} u_j^\sigma \quad \forall \sigma \in \Pi$$
$$\pi_j + u_j^\sigma + w_S^\sigma \geq v_j(S) \quad \forall j \in S, \forall S \in \sigma$$
$$\pi_j, u_j^\sigma, w_S^\sigma, \pi^s \geq 0 \quad \forall j \in N, \forall S \in \sigma, \forall \sigma \in \Pi$$

3. Heterogeneous goods, unit demand. The first special case we consider is when each agent is interested in at most one object. It is well known that in this case the "agents are substitutes" condition (1) holds. Following formulation (P4), this means that there is only one partition of the set M of objects that needs to be considered. (P4) thus reduces to an assignment problem.

To fix notation, let $v_{ij} \geq 0$ be the value that agent $j \in N$ assigns to object $i \in M$. By adding dummy objects of zero value to all agents, we can always ensure that $|M| \geq |N|$. To formulate the problem of finding an efficient allocation let $x_{ij} = 1$ if agent j is allocated object i and zero otherwise.

$$\max \sum_{j \in N} \sum_{i \in M} v_{ij} x_{ij}$$
$$\text{s.t. } \sum_{j \in N} x_{ij} \leq 1 \quad \forall i \in M$$
$$\sum_{i \in M} x_{ij} \leq 1 \quad \forall j \in N$$
$$x_{ij} \geq 0$$

This, of course, is the well known assignment problem in which the linear program has all integral extreme points. The dual to it is:

$$\min \sum_{j \in N} u_j + \sum_{i \in M} w_i$$
$$\text{s.t. } u_j + w_i \geq v_{ij} \quad \forall j \in N, \forall i \in M$$
$$u_j, w_i \geq 0 \quad \forall j \in N, \forall i \in M$$

In Leonard (1983) it is shown that among all optimal dual solutions, the one that maximizes $\sum_{j \in N} u_j$ yields the Vickrey payments. Specifically u_j is the marginal product of agent j. If one thinks about the definition of dual variables as the change in objective function value from a small change in the right hand side of the relevant constraint, then this should come as no surprise in the unit demand setting. Outside this setting, it need not be true. If we interpret w_i as the price of object i, we see that the price an agent pays for the object is anonymous; that is, it depends on the object alone. Given the form of (DP4), this is to be expected.

More interesting is that a particular implementation of the primal-dual algorithm for solving the assignment problem—one which finds, among optimal dual solutions, the one that maximizes total bidder surplus—produces an ascending auction that duplicates the outcome of the sealed bid Vickrey auction. Further, it is a Nash equilibrium for players to bid truthfully in each round of the auction. Such an auction was first proposed by Crawford and Knoer (1981) under the assumption that bidders have strict preferences (that is, no bidder is indifferent between any two objects). When the strict preference condition is dropped, the Crawford and Knoer auction/algorithm can be made to produce an outcome arbitrarily close to the optimal solution of the assignment by choosing a tie-breaking parameter to be sufficiently small. Here we describe a version of the Crawford–Knoer algorithm due to Demange, Gale and Sotomayor (1986) that sidesteps the tie-breaking issues.

First we review the primal-dual algorithm for the assignment problem. Fix nonnegative $(w_i)_{i\in M} \geq 0$ and choose u so that $u_j = \max_i[v_{ij} - w_i]^+$ for $j \in N$. The resulting choice produces a feasible dual solution. We can interpret w as the price vector of objects and u as the vector of maximum profits that bidders achieve at the current prices. For each agent j let $D_j(w) = \{i : [v_{ij} - w_i]^+ = u_j\}$. Thus $D_j(w)$ is the set of objects that maximize agent j's payoff at the current prices w.

Assuming the current feasible dual solution were optimal, we could determine a primal solution by enforcing the complementary slackness conditions. Thus an agent j and an object i such that $i \notin D_j(w)$, i.e., $u_j + w_i > v_{ij}$ would require that $x_{ij} = 0$. Hence the only variables, x_{ij} in the primal that can be set to 1 are those for which $u_j + w_i = v_{ij}$, i.e., $i \in D_j(w)$. The next step is to use only these variables to assign as many objects to agents so that no agent gets more than one object and no object is assigned to more than one agent. This auxiliary optimization problem, sometimes called the restricted primal **(RP)**, is:

$$\max \sum_{j\in N} \sum_{i\in D_j(w)} x_{ij}$$

$$\text{s.t.} \quad \sum_{j\in N:\, i\in D_j(w)} x_{ij} \leq 1 \quad \forall i \in M$$

$$\sum_{i\in D_j(w)} x_{ij} \leq 1 \quad \forall j \in N$$

$$x_{ij} \geq 0$$

The optimal solution to the restricted primal will either find an assignment in which all agents receive exactly one object or prove that no such assignment exists. In the first case, stop; the optimal solution has been found. In the second case, when no such assignment exists (assuming the appropriate solution algorithm is used), a set S of objects such that $|S| < |\{j : D_j(w) \subseteq S\}|$ will be found. The set S is called an **overdemanded set.** The existence of such a set is proven by Demange, Gale, and Sotomayor (1986) via Hall's Marriage Theorem. It can be derived from the dual of the restricted primal.

The optimal dual solution of the restricted primal determines the "direction" in which the current dual solution should be adjusted. If (u, w) is the current dual solution and (u', w') is the optimal solution to the dual of the restricted primal, then the adjusted dual solution for the next iteration is $(u, w) + \theta(u', w')$, where $\theta > 0$ is a user-specified step size. In the case of the assignment problem, each $w_i \in S$ is increased by $\theta > 0$ (suitably chosen) and each u_j is then recomputed.

Different primal-dual algorithms differ by the choice of the overdemanded set used and how the dual variables are adjusted. Since we want the dual variables to yield the Vickrey payoffs, we need to ensure that a particular optimal dual solution is produced. We need the optimal dual solution with the largest $\sum_{j \in N} u_j$ or, equivalently, smallest $\sum_{i \in M} w_i$. This can be done by choosing, at each iteration, a *minimal* overdemanded set.

Since such algorithms will appear in subsequent sections, it will be useful to provide a high level description. Algorithm 1 is from Papadimitriou and Steiglitz (1982).

Algorithm 1 The Primal-Dual Algorithm (Papadimitriou & Steiglitz).

```
1: infeasible←'no'
2: opt←'no'
3: start with a dual feasible solution
4: while infeasible= 'no' and opt='no' do
5:    J←indices of dual inequalities fulfilled with equality
6:    solve the restricted primal (RP)
7:    if opt of (RP) = 0 then
8:       opt←'yes'
9:    else
10:      solve the dual of the restricted primal
11:      if the product of the solution with each column of the primal
         constraint matrix is nonpositive then
12:         infeasible←'yes'
13:      else
14:         update the dual solution
15:      end if
16:   end if
17: end while
```

We now describe the auction. Suppose that bidders valuations are integral.[5] The auction proceeds in rounds. At the start of round t, let p_i^t be the current price of object i. We set $p_i^0 = 0$ for all $i \in M$. At the beginning of the round each bidder is asked to name all their 'favorite objects', i.e., the set $\arg\max_i[v_{ij} - p_i^t]$. Bidders whose favorite objects yield negative payoff leave the auction.

Let B^t be a bipartite graph with a node for each remaining bidder and one for each object. Insert edges between a bidder and an object that is her favorite. If the graph has a matching such that every bidder is assigned to exactly one object and no object is assigned to more than one bidder, stop. The assignment is made at the price p^t. If no such matching exists, there is, by Hall's Marriage theorem, a subset B of bidders such that $|B| > |N(B)|$ where $N(B)$ is the set of neighbors of B. The set $N(B)$ is called an overdemanded set of objects. From among all overdemanded sets, choose one that is minimal. Raise the price on all objects in that set by \$1, terminating round t of the auction.

It is clear that if bidders bid truthfully in each round, the auction will mimic the primal-dual algorithm for the assignment problem. In fact, it is a Nash equilibrium for the bidders to bid truthfully in each round. The intuition for this fact can be described as follows. Since all other bidders are bidding truthfully, the only thing a bidder can do in each round is to prevent prices from increasing. Prices rise only on objects in a minimally overdemanded set. Thus, to affect a price rise, an agent's favorite objects (at the current price) must lie in this overdemanded set. The only way the agent can prevent prices rising on objects in this set is by reducing his own demand for objects in this set, thereby making the agent worse-off.

4. Heterogeneous goods, gross substitutes. Gul and Stacchetti (2000) have proposed an algorithm that generalizes the auction procedure of Demange, Gale, and Sotomayor (1986) to environments in which bidders have preferences over subsets of items. It is a primal-dual algorithm, but does not share the incentive properties of the Demange, Gale, and Sotomayor algorithm. It also is a variation of the Kelso and Crawford (1982) algorithm that sidesteps tie-breaking issues when agents are indifferent between utility maximizing bundles of objects.

In this set-up, agents have preferences over subsets of heterogeneous objects. The value that bidder j assigns to the set $S \subset M$ of objects is $v_j(S)$. An efficient allocation is a partition σ of the objects among the agents that maximizes the sum of values, i.e., problem (P1).

As in Kelso and Crawford (1982), Gul and Stacchetti (2000) impose two conditions on agent's value functions to ensure that the linear relaxation of (P1) has an optimal integer solution. They then provide a primal-dual algorithm for the optimization problem and give it an auction interpretation. To describe these two properties we need some notation.

[5] A standard scaling argument will take care of the non-integral case.

Given a vector of prices p on objects let the collection of subsets of objects that maximize bidder j's utility be denoted $D_j(p)$, defined as follows.

$$D_j(p) = \{S \subset M : v_j(S) - \sum_{i \in S} p_i \geq v_j(T) - \sum_{i \in T} p_i \quad \forall T \subset M\}.$$

Monotonicity: For all $j \in N$ and all $A \subset B \subset M$, $v_j(A) \leq v_j(B)$.
Single improvement: For all $A \notin D_j(p)$ there exists $B \subset M$ such that $v_j(B) - \sum_{i \in B} p_i > v_j(A) - \sum_{i \in A} p_i$ and $|A \setminus B|, |B \setminus A| \leq 1$.
Together, these two conditions imply that v_j is submodular.

While monotonicity seems to be economically plausible, the single improvement property does not. In Gul and Stacchetti (1999), however, it is shown that the single improvement property (under monotonicity) is equivalent to a property called *gross substitutes*, which does have economic content. The gross substitutes property, introduced by Kelso and Crawford (1982), requires that a demanded object still be demanded if the prices of all other objects increase.

Gross substitutes: For all price vectors p, p' such that $p' \geq p$, and all $A \in D_j(p)$, there exists $B \in D_j(p')$ such that $\{i \in A : p_i = p'_i\} \subset B$.

These two conditions impose two useful properties on the sets $D_j(p)$. The first is that

$$r_j(T, p) = \min_{S \in D_j(p)} |S \cap T|$$

is the dual rank function of a matroid. From this it follows that for any $T \subset M$ and pair of price vectors $p' \geq p$ such that $p_i = p'_i$ for all $i \in M \setminus T$, we have

$$r_j(T, p) \leq r_j(T, p').$$

Proofs of both statements can be found in Gul and Stacchetti (2000).

Recall formulation (P1):

$$\begin{aligned} V(N) = \max & \sum_{j \in N} \sum_{S \subseteq M} v_j(S) y(S, j) \\ \text{s.t.} & \sum_{S \ni i} \sum_{j \in N} y(S, j) \leq 1 \quad \forall i \in M \\ & \sum_{S \subseteq M} y(S, j) \leq 1 \quad \forall j \in N \\ & y(S, j) = 0, 1 \quad \forall S \subseteq M, \forall j \in N \end{aligned}$$

Relaxing the integrality constraint in (P1) gives

$$V_{LP}(N) = \max \sum_{j \in N} \sum_{S \subseteq M} v_j(S) y(S,j)$$
$$\text{s.t.} \sum_{S \ni i} \sum_{j \in N} y(S,j) \leq 1 \quad \forall i \in M$$
$$\sum_{S \subseteq M} y(S,j) \leq 1 \quad \forall j \in N$$
$$y(S,j) \geq 0 \quad \forall S \subseteq M, \forall j \in N$$

THEOREM 4. *When each agent's value function satisfies monotonicity and the single improvement property,* $V(N) = V_{LP}(N)$.

Proof. Let p^* be an optimal dual solution to the linear relaxation of (P1). Let

$$V_p(N) = \max \sum_{j \in N} \sum_{S \subseteq M} v_j(S) y(S,j) - \sum_{i \in M} p_i \left[\sum_{S \ni i} \sum_{j \in N} y(S,j) \right] + \sum_{i \in M} p_i$$
$$\text{s.t.} \sum_{S \subseteq M} y(S,j) \leq 1 \quad \forall j \in N$$
$$y(S,j) \geq 0 \quad \forall S \subseteq M, \forall j \in N$$

Rewriting the objective function,

$$V_p(N) = \max \sum_{j \in N} \sum_{S \subset M} \left[v_j(S) - \sum_{i \in S} p_i \right] y(S,j) + \sum_{i \in M} p_i$$
$$\text{s.t.} \sum_{S \subseteq M} y(S,j) \leq 1 \quad \forall j \in N$$
$$y(S,j) \geq 0 \quad \forall S \subseteq M, \forall j \in N$$

In this problem, if $y(S,j) = 1$ is part of an optimal solution, then $S \in D_j(p)$. By the duality theorem of linear programming, $V_{p^*}(N) = V_{LP}(N)$. If there is a partition of M so that each agent j receives at most one element of $D_j(p)$, it would follow that (P1) has an optimal integer solution. Suppose no such partition exists. Given that each r_j is a dual rank function of a matroid, it follows from the matroid partition theorem (Edmonds and Fulkerson (1965)) that there is a $T \subset M$ such that

$$\sum_{j \in N} r_j(T, p^*) > |T|.$$

Choose a new dual solution p such that $p_i = p_i^*$ for all $i \in M \setminus T$ and $p_i = p_i^* + \epsilon$ for all $i \in T$ for sufficiently small $\epsilon > 0$. For each $j \in N$, choose $B^j \subset M$ such that $r_j(T, p^*) = |B^j \cap T|$. Then

$$\begin{aligned}
V_p(N) &= \sum_{j\in N}\left[v_j(B^j) - \sum_{i\in M} p_i\right] + \sum_{i\in M} p_i^* + \epsilon|T| \\
&= \sum_{j\in N}\left[v_j(B^j) - \sum_{i\in M} p_i^*\right] - \epsilon\sum_{j\in N}|B^j\cap T| + \sum_{i\in M} p_i^* + \epsilon|T| \\
&= V_{p^*}(N) + \epsilon|T| - \epsilon\sum_{j\in N} r_j(T,p^*) \\
&= V_{p^*} - \epsilon(\sum_{j\in N} r_j(T,p^*) - |T|) < V_{p^*}(N)
\end{aligned}$$

which contradicts the optimality of p^*. □

The proof suggests a primal-dual algorithm. Fix a price vector p. At those prices, determine a partition of M that gives to each agent an element of $D_j(p)$. If no such partition exists, find an overdemanded set T, i.e., a set T such that

$$\sum_{j\in N} r_j(T,p^*) > |T|.$$

Particularly, choose a minimal one, according to the algorithm proposed by Gul and Stacchetti (2000), and raise the price on each object in that set by $\epsilon > 0$. Assuming sincere bidding at each round, the auction produces, at termination, *minimal* Walrasian prices.[6] As noted by Gul and Stacchetti, the auction is not guaranteed to produce Vickrey prices. We now provide an example to demonstrate.

The example consists of three agents, each of whom may consume at most two of four given objects. Their valuations for such subsets are additive. Let $N = \{1,2,3\}$ and $M = \{a,b,c,d\}$. The agents value single-object subsets as shown in Table 1. For $|S| = 2$, where $\{i,i'\} = S$, let $v_j(S) = v_j(i) + v_j(i')$. For $|S| > 2$, let $v_j(S) = \max\{v_j(S') : S' \subset S, |S'| = 2\}$.

TABLE 1
$v_j(S)$ *when* $|S| = 1$.

	a	b	c	d
1	10	10	10	10
2	4	7	9	8
3	2	2	2	4

[6]Prices are Walrasian for an allocation of objects if, given those prices, each agent is consuming a set of objects that maximizes his payoff (net of prices). See any standard microeconomics textbook for a formal definition, and see the example below.

The efficient allocation is to give objects a and b to agent 1, and the other two objects to agent 2, i.e., $y(\{a,b\},1) = 1$ and $y(\{c,d\},2) = 1$. For (p_a, p_b, p_c, p_d) to be a *Walrasian price vector* supporting this allocation, each agent must be consuming his most preferred bundle of objects, given these prices.

This requires many inequalities to hold. In particular, since agent 1 is consuming b and not c, this requires $p_b \leq p_c$. Since agent 2 consumes d and not b, this requires $p_d \leq p_b + 1$. Finally, since agent 3 consumes no object, this requires $p_a \geq 2$ and $p_d \geq 4$.

Therefore, the minimal Walrasian price vector[7] is $(2,3,3,4)$. However, agent 2's Vickrey payment is $6 < p_c + p_d = 7$. Therefore, in this example, the agents' Vickrey payments cannot be additively decomposed into a vector of Walrasian prices for the objects.

Recently, Parkes and Ungar (2000) have proposed a primal-dual algorithm with an auction interpretation for (P3). This allows them to drop the gross substitutes assumption and determine non-linear, non-anonymous prices that support an efficient allocation. These are not, however, Vickrey prices.

4.1. Agents are substitutes. Here we show that when all agents' value functions are monotonic and satisfy the single improvement property, the substitutes condition (1) holds. This condition enables us to invoke formulation (P4) to derive the Vickrey prices. From (DP4) we see that the Vickrey prices cannot be anonymous.

For every subset K of agents, let

$$\begin{aligned} V_p(K) = \max &\sum_{j \in K} \sum_{S \subset M} (v_j(S) - \sum_{i \in S} p_i) y(S,j) + \sum_{i \in M} p_i \\ \text{s.t.} &\sum_{S \subseteq M} y(S,j) \leq 1 \quad \forall j \in K \\ & y(S,j) \geq 0 \quad \forall S \subseteq M, \forall j \in K. \end{aligned}$$

$V_p(K)$ can be expressed more succinctly as:

$$V_p(K) = \sum_{j \in K} \max_{S \subseteq M} [v_j(S) - \sum_{i \in S} p_i] + \sum_{i \in M} p_i$$

By the duality theorem of linear programming:

$$V(K) = \min_{p \geq 0} V_p(K).$$

Amongst all $p \in \arg\min_{p \geq 0} V_p(K)$ let p^K denote the minimal one. From Gul and Stacchetti (1999) we know that the set $\arg\min_{p \geq 0} V_p(K)$ forms a

[7]The vector is minimal in the sense that every other Walrasian price vector weakly dominates this one.

complete lattice and a minimal element exists. It is also easy to see that if $K \subset T$ then $p^K \leq p^T$. Associated with each $V(K)$ is a partition of M. Denote by S_j^K the subset of objects that agent j receives in this partition, i.e., $y(S_j^K, j) = 1$.

For each $k \in N$ we have

$$V(N \setminus k) = V_{p^{N\setminus k}}(N \setminus k) \geq \sum_{j\in N\setminus k} [v_j(S_j^N) - \sum_{i\in S_j^N} p_i^{N\setminus k}] + \sum_{i\in M} p_i^{N\setminus k}$$

$$= \sum_{j\in N\setminus k} v_j(S_j^N) + \sum_{i\in S_k^N} p_i^{N\setminus k}$$

Hence

$$V(N) - V(N \setminus k) \leq \sum_{j\in N} v_j(S_j^N) - \sum_{j\in N\setminus k} v_j(S_j^N) - \sum_{i\in S_k^N} p_i^{N\setminus k}$$

$$= v_k(S_k^N) - \sum_{i\in S_k^N} p_i^{N\setminus k}.$$

Let p' be a price vector such that

$$p'_i = p_i^N \quad \forall i \in \bigcup_{j\in K} S_j^N$$

and

$$p'_i = p_i^{N\setminus k} \quad \forall i \in S_k^N, \forall k \in N \setminus K.$$

Then

$$V(K) = V_{p^K}(K) \leq V_{p'}(K) = \sum_{j\in K} \max_{S\subseteq M}[v_j(S) - \sum_{i\in S} p_i^N] + \sum_{i\in M} p'_i$$

$$= \sum_{j\in K} v_j(S_j^N) + \sum_{k\in N\setminus K} \sum_{i\in S_j^N} p_i^{N\setminus k}$$

Therefore,

$$V(N) - V(K) \geq \sum_{j\in N\setminus K} v_j(S_j^N) - \sum_{k\in N\setminus K} \sum_{i\in S_j^N} p_i^{N\setminus k} \geq \sum_{k\in N\setminus K} [V(N) - V(N\setminus k)].$$

5. Spanning trees. Let $G = (V, E)$ be a connected graph with vertex set V and edge set E. Each edge of the graph is owned by some agent. Denote by E_j the set of edges owned by agent $j \in N$, and let $E_K = \bigcup_{j\in K} E_j$. The cost of using an edge $e \in E$ is c_e. An efficient allocation is a set of edges that span the graph with minimal total cost.

Let T denote the minimum spanning tree (MST) and let L denote its length. For any subset $E' \subset E$ of edges, let $T^{-E'} \subset E \setminus E'$ be the MST that does not use any edge in E'; denote its length by $L^{-E'}$, which by convention is equal to ∞ if there is no spanning tree after deleting edges in E'.

For the remainder of our discussion on spanning trees, we assume that no agent owns a cut; that is, for each agent $i \in N$, we assume that $L^{-E_i} < \infty$. We show that with this assumption in this model, agents are substitutes.

THEOREM 5. *In the minimum spanning tree problem, the "agents are substitutes" condition (1) holds (when no agent owns a cut): For all* $K \subset N$,

$$L^{-E_K} - L \geq \sum_{i \in K} [L^{-E_i} - L].$$

Proof. A sufficient condition for the "agents are substitutes" condition to hold is that an agent's marginal contribution to coalitions is monotonically decreasing as the coalition increases. In other words, for the MST problem, it is sufficient to show that for all $K \subset K' \subseteq N$ and all $i \notin K'$,

$$L^{-E_{K \cup i}} - L^{-E_K} \leq L^{-E_{K' \cup i}} - L^{-E_{K'}}.$$

By applying this monotonicity condition to a telescoping sum, the substitutes condition is proven.

In turn, the following condition, which we prove, implies the above monotonicity condition.

$$\forall a \neq e \in E, \quad L^{-a,e} - L^{-e} \geq L^{-a} - L$$

This condition says that as other edges are removed from a graph, the marginal value of a given edge increases.

If either $L^{-a} = L$ or $L^{-e} = L$, we are done. Otherwise, there exist $b, b', f, f' \in E$ such that

$$\begin{aligned} T^{-a} &= (T \setminus a) \cup b & \qquad T^{-e} &= (T \setminus e) \cup f \\ T^{-ae} &= (T^{-e} \setminus a) \cup b' & \qquad T^{-ae} &= (T^{-a} \setminus e) \cup f' \end{aligned}$$

Note that $\{b, b', f, f'\} \subset T^{-ae}$. Therefore, if $b \neq f$, then $T^{-ae} = T \setminus \{a, e\} \cup \{b, f\}$, and $L^{-a,e} - L^{-e} = L^{-a} - L$. Otherwise, if $b = f$, then $b \in T^{-e}$, so $b' \neq b$. Similarly, $f' \neq f$. Therefore $b' = f'$.

What we are trying to show from the above inequality is

$$(L^{-a} - c_e + c_{f'}) - (L - c_e + c_f) \geq L^{-a} - L$$

which is equivalent to $c_{f'} \geq c_f$. Since $c_b = c_f$ and $c_{b'} = c_{f'}$, it is clear that $c_{f'} \geq c_f$, otherwise either T^{-a} or T^{-e} would not have been minimal. □

5.1. Agents own one edge. There are many ways to formulate the problem of finding a minimum cost spanning tree as a linear program. In fact we could use formulations (P3) or (P4) directly. Instead we use a standard "covering" formulation as in Bertsimas and Teo (1998) for the case when $|E_j| = 1$. This allows us to extend the arguments made here to the more general case of matroids.

Let Π be the set of partitions of V. For any partition $\sigma \in \Pi$, $S \in \sigma$ means that $S \subset V$ is a part of the partition σ and $|\sigma|$ refers to the number of parts of this partition. For each partition of vertices $\sigma \in \Pi$, let $\delta(\sigma) \subset E$ be the set of edges with endpoints in different parts of σ. Set $x_e = 1$ if edge $e \in E$ is selected and zero otherwise.

$$\min \sum_{e \in E} c_e x_e$$

$$\text{s.t.} \sum_{e \in \delta(\sigma)} x_e \geq |\sigma| - 1 \quad \forall \sigma \in \Pi$$

$$\sum_{e \in E} x_e = |V| - 1$$

$$0 \leq x_e \leq 1 \quad \forall e \in E$$

Actually, the constraint $x_e \leq 1$ is redundant, but its dual will yield agent e's marginal product. It is well known that this formulation is integral.

Let λ_e be the dual variable associated with the constraint $x_e \leq 1$, and μ_σ the dual variable associated with $\sum_{e \in \delta(\sigma)} x_e \geq |\sigma| - 1$. Let μ be the dual variable associated with $\sum_{e \in E} x_e = |V| - 1$. The dual problem is

$$\max - \sum_{e \in E} \lambda_e + (|V| - 1)\mu + \sum_{\sigma \in \Pi} (|\sigma| - 1)\mu_\sigma$$

$$\text{s.t.} \ -\lambda_e + \mu + \sum_{\delta(\sigma) \ni e} \mu_\sigma \leq c_e \quad \forall e \in E$$

$$\lambda_e \geq 0 \quad \forall e \in E$$

$$\mu_\sigma \geq 0 \quad \forall \sigma \in \Pi$$

As before let T be the minimum spanning tree. If $e \notin T$ set $\lambda_e = 0$. If $e \in T$ set $\lambda_e = c_{f_e} - c_e$ where f_e is the shortest edge such that $T \cup f_e$ contains a cycle through e. Clearly, each λ_e is agent e's marginal product. For this choice of λ_e we must show that we can choose μ and μ_σ to be dual feasible and to yield an objective function value equal to the length of the minimum spanning tree.

Substituting in the value of λ_e we have specified into the dual and dropping the part $-\sum_{e \in E} \lambda_e$ from the objective function yields

$$\max\,(|V|-1)\mu + \sum_{\sigma\in\Pi}(|\sigma|-1)\mu_\sigma$$
$$\text{s.t. } \mu + \sum_{\delta(\sigma)\ni e}\mu_\sigma \le c_e \quad \forall e \notin T$$
$$\mu + \sum_{\delta(\sigma)\ni e}\mu_\sigma \le c_{f_e} \quad \forall e \in T$$
$$\mu_\sigma \ge 0 \quad \forall \sigma \in \Pi.$$

This is the dual to the problem of finding a minimum spanning tree in the original graph whose edge weights have been modified. Each edge $e \in T$ has had its weight increased to c_{f_e}. The length of the minimum spanning tree in this modified graph is of course $\sum_{e\in T} c_{f_e}$.

Hence the optimal values of μ and μ_σ from the second dual program combined with our choice of λ_e form a feasible solution to the original dual problem. The value of this dual solution is:

$$-\sum_{e\in T}(c_{f_e} - c_e) + \sum_{e\in T} c_{f_e} = \sum_{e\in T} c_e.$$

That is, it coincides with the optimal value of the primal.

A primal-dual algorithm for this formulation requires that at each iteration, prices on $\delta(\sigma)$ be announced. These can be interpreted as prices for inserting an edge connecting different parts of σ. Roughly speaking, the auctioneer runs a second price auction on each partition of vertices. This observation allows us to derive from the primal-dual algorithm a descending auction (since we are in a procurement setting) to implement the Vickrey outcome. We omit a description of the primal-dual algorithm and proceed directly to the auction.

To run the auction, we first set the offered price on each edge at some large number P. Each agent is to announce whether they are willing to offer their edge at that price. Now, we begin to drop the price. As the price drops, agents withdraw their edges when they become unprofitable at the given price, reducing the connectivity of the graph. When the connectivity of the graph drops to 1, the auction pauses, and the agent who owns the critical edge is given the price at which the auction has paused for that edge. That edge is not removed for the duration of the auction, and is committed to being a part of the final tree. Now restart the auction, pausing at the next critical edge whose removal would disconnect the graph, and so on, making a payment and committing an edge at each pause. The auction ends when a tree has been completed.

5.2. Agents own multiple edges. We now give a formulation for the case in which $|E_j| \ge 1$. For each $F \subseteq E_j$, set $c_j(F) = \sum_{e\in F} c_e$. For each $j \in N$ and $F \subseteq E_j$, set $y_j(F) = 1$ if F is selected to be in a MST and zero otherwise. The formulation, which we denote **(TP)**, is

$$
\begin{aligned}
\min & \sum_{j \in N} \sum_{F \subseteq E_j} c_j(F) y_j(F) \\
\text{s.t.} & \sum_{j \in N} \sum_{F \subseteq E_j :\, \delta(\sigma) \cap F \neq \emptyset} y_j(F) \geq |\sigma| - 1 \quad \forall \sigma \in \Pi \\
& \sum_{j \in N} \sum_{F \subseteq E_j} |F| y_j(F) = |V| - 1 \\
& - \sum_{F \subset E_j} y_j(F) \geq -1 \quad \forall j \in N \\
& y_j(F) \geq 0 \quad \forall j \in N, \forall F \subseteq E_j.
\end{aligned}
$$

The first constraint ensures that the tree spans the graph; the second says that exactly $|V| - 1$ edges are used (so this could be weakened to an inequality); the third says that each agent supplies at most one subset of edges, F.

We will show that this formulation has an optimal integral solution. In addition, the dual variable corresponding to the constraint $-\sum_{j \in N} y_j(F) \geq -1$ yields agent j's marginal product.

THEOREM 6. *Formulation (TP) has an optimal integral solution.*

Proof. Let y denote an optimal extreme (possibly fractional) solution to (TP). For each agent j and edge $e \in E_j$, set

$$
z_e = \sum_{F \subseteq E_j} y_j(F).
$$

We will show that z is a feasible solution to the following system:

$$
\begin{aligned}
& \sum_{e \in \delta(\sigma)} x_e \geq |\sigma| - 1 \quad \forall \sigma \in \Pi \\
& \sum_{e \in E} x_e = |V| - 1 \\
& 0 \leq x_e \quad \forall e \in E
\end{aligned}
$$

which we know to have integral extreme points. Furthermore, since we will prove that each vertex of the latter formulation has an integral preimage in the previous formulation, we conclude that the latter formulation is a projection of the earlier formulation. Given an integral solution x^* to this system, we can construct an integral solution y^* to (TP). Let $G_j = \{e \in E_j : x_e^* = 1\}$. For all $j \in N$, set $y_j^*(G_j) = 1$ and $y_j^*(F) = 0$ for $F \neq G_j$. Observe that $\sum_e c_e x_e^* = \sum_{j \in N} c_j(G_j) y_j(G_j)$, i.e. both solutions have the same objective function value.

Clearly $z_e \geq 0$. Also,

$$
\sum_e z_e = \sum_{j \in N} \sum_{e \in E_j} \sum_{F \subseteq E_j : e \in F} y_j(F) = \sum_{j \in N} \sum_{F \subseteq E_j} |F| y_j(F) = |V| - 1.
$$

Next,

$$\sum_{e\in\delta(\sigma)} z_e = \sum_{j\in N} \sum_{e\in E_j\cap\delta(\sigma)} \sum_{F\subseteq E_j: e\in F} y_j(F)$$
$$= \sum_{j\in N} \sum_{F\subseteq E_j: F\cap\delta(\sigma)\neq\emptyset} y_j(F) \geq |\sigma| - 1.$$

Notice also that

$$\sum_{j\in N} \sum_{F\subseteq E_j} c_j(F) y_j(F) = \sum_e c_e z_e.$$

Hence, if z_e is integral we are done. If not, we can express z_e as a convex combination of integral points of the second formulation. But each of these corresponds in turn to an integral point of (TP). Thus, y can be expressed as a convex combination of extreme points of (TP), a contradiction. □

The dual to the formulation (TP) is

$$\max \sum_{S\subset V} \mu_S + (|V|-1)\mu - \sum_{j\in N} \lambda_j$$
$$\text{s.t.} \quad \sum_{F\subseteq E_j:\, \delta(S)\cap F\neq\emptyset} |\delta(S)\cap F|\mu_S + |F|\mu - \lambda_j \leq c_j(F) \quad \forall F\subseteq E_j$$
$$\mu_S \geq 0 \quad \forall S\subset V$$
$$\lambda_j \geq 0 \quad \forall j\in N.$$

As before, T denotes the MST, and T^{-E_j} is the MST subject to the constraint that no edges in E_j are used. Let $F^j = T\cap E_j$ be the set of edges owned by agent j in a MST. Observe that $|F^j| = |T^{-E_j}\setminus T|$, i.e., when edges are deleted from possible use, they are replaced on a one-for-one basis. Let

$$\lambda_j = \left(\sum_{e\in T^{-E_j}\setminus T} c_e\right) - c_j(F^j).$$

Notice this is agent j's marginal product.

Order the edges in F^j by increasing cost: $e_1, e_2, \ldots, e_t$ for some t. Do the same with edges in $T^{-E_j}\setminus T$: $e'_1, e'_2, \ldots, e'_t$. It is easy to check that $c_{e_i} \leq c_{e'_i}$ for $1\leq i\leq t$. Also $T\cup e'_i$ has a cycle that goes through e_i and e'_i. Suppose we modify the costs of the edges by increasing the cost of each e_i to $c_{e'_i}$. Then T will still be a minimum spanning tree with respect to the modified weights. Let c' denote the vector of modified costs.

Substituting the value for λ into the dual yields

$$-\sum_{j\in N}\lambda_j + \max\sum_{\sigma\in\Pi}\mu_\sigma + (|V|-1)\mu$$

$$\sum_{\sigma:\delta(\sigma)\cap F\neq\emptyset} |\delta(\sigma)\cap F|\mu_\sigma + |F|\mu \leq \lambda_j + c_j(F) \quad \forall F\subseteq E_j$$

$$\mu_\sigma \geq 0 \quad \forall\sigma\in\Pi$$

It suffices to show that we can find a feasible solution to the dual with objective function value $\sum_{e\in T} c_e$. By the previous theorem this dual solution must be optimal.

Notice that any feasible solution to

$$\sum_{\sigma:\delta(\sigma)\cap F\neq\emptyset} |\delta(\sigma)\cap F|\mu_{sigma} + |F|\mu \leq c'_j(F) \quad \forall F\subseteq E_j$$

$$\mu_\sigma \geq 0 \quad \forall\sigma\in\Pi$$

is a feasible solution to the dual because

$$\lambda_j + c_j(F) = \left(\sum_{e\in T^{-E_j}\setminus T} c_e\right) - c_j(F^j) + c_j(F)$$

$$= \left(\sum_{e\in T^{-E_j}\setminus T} c_e\right) - c_j(F^j\setminus F) + c_j(F\setminus F^j) \geq c'(F).$$

However, this second system is the dual to the minimum spanning tree problem with respect to the cost vector c'. Thus this second system admits a feasible solution with objective function value $\sum_{e\in T} c'_e = \sum_{j\in N}\sum_{e\in F^{-j}} c_e$. Substituting these variables into the original dual system gives us a solution with objective function value

$$-\sum_{j\in N}\lambda_j + \sum_{j\in N}\sum_{e\in F^{-j}} c_e = \sum_{e\in T} c_e.$$

By the previous theorem this dual solution must be optimal.

5.3. Matroids. The problem of finding a minimum weight spanning tree is a special case of the problem of finding a minimum/maximum weight basis in a matroid.[8] Here we outline how the descending auction can easily be adapted to the matroid case.

Let E be the ground set and E^r the subset of E owned by agent r. For each $e\in E$ let w_e be the weight of e known only to the agent who owns

[8]This section assumes familiarity with matroid theory. Gul and Stacchetti (2000) also make an appeal to matroid theory.

that element. Let I be the family of independent subsets of E. The pair (E, I) forms a matroid.

Now we describe the auction. Start with a high 'offer' price for all elements. Agents announce which elements they are prepared to offer at the current price. These are called active elements. Now drop the price. As the price drops, agents withdraw some elements. Meanwhile, the auctioneer checks that for any agent, a basis is contained in the set of active elements not belonging to that agent.

Consider the instant this condition is violated. Let j be the agent, p the current price and E_a^j the currently active set of elements belonging to agent j. Let M be any minimal subset of E_a^j whose removal from the set of all active edges would eliminate the existence of a basis. From the matroid property, all such sets have the same cardinality. The auctioneer pays $|M|p$ to agent j, and agent j chooses which subset of size $|M|$ from E_a^j to proffer. The price on these elements is fixed at p for the remainder of the auction, and the auction continues.

If the goal is to find a maximum weight basis, then start with a low price and raise it. The rest of the auction works as before.

Consider now an auction where K identical units must be auctioned off to n bidders. The value that bidder j assigns to their i^{th} unit is v_j^i. We assume that $v_j^i \geq v_j^{i+1}$, indicating that bidders have diminishing marginal utilities. The problem of finding an efficient allocation can be formulated as the problem of finding a maximum weight basis.

For each bidder j we introduce K elements, $j^1, j^2, \ldots, j^K$ with weights $v_j^1, v_j^2, \ldots, v_j^K$. Let E be the collection of all such elements. Let I be the collection of all subsets of size at most K. It is easy to see that (E, I) is a matroid. In this case the ascending auction coincides with Ausubel's (1997) ascending auction for homogeneous products.

6. Shortest paths. Let $G = (V, E)$ be a directed graph with vertex set V and edge set E. Specify a source node $s \in V$ and sink node $t \in V$. Each edge $e \in E$ has a cost c_e. In what follows, we sometimes refer to an edge e as a pair of vertices (i, j) where e is directed from i to j. The efficient allocation is the s–t path of minimal total cost, i.e., the shortest path.

Using the results of Hoffman and Markowitz (1963), we can convert the shortest path problem into an assignment problem. Replace each vertex $i \in V$ by two vertices i and i'. Replace each directed edge (i, j) by two directed edges: (i, j') with cost c_{ij}, and (i, i') with cost zero. Delete vertices s' and t from this extended network. An optimal assignment in this new graph corresponds to a shortest path in the original one, and vice versa.

Given this correspondence, it is tempting to think that one can apply the methods of Demange, Gale, and Sotomayor (1986) to this problem; unfortunately one cannot. Even if each agent is associated with only a single edge, each agent appears on both sides of this assignment problem.

To be precise about the ownership structure, each agent j owns a set of edges E^j. The interesting cases are those in which the following no-monopoly condition holds, which we now assume: no agent owns a cut that disconnects s from t.

In this model, the substitutes condition (1) need not hold. To illustrate the complementarities of agents, we give an example even for the case in which each s–t path uses at most one edge belonging to any particular agent. The example involves three edges and three agents. To each directed edge in Figure 1 is associated an agent–cost pair (e.g., agent A's edge has a cost of 1).

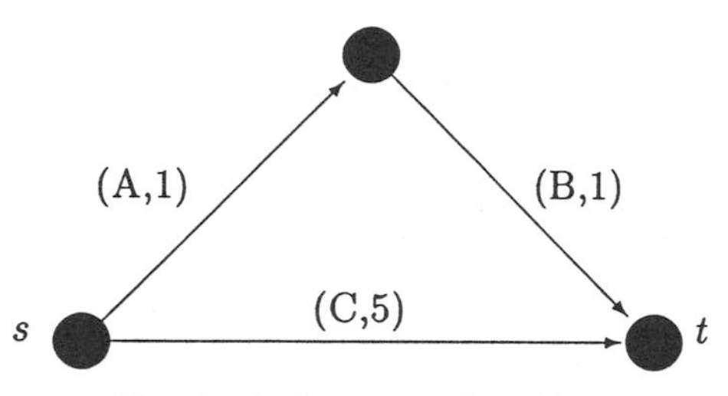

FIG. 1. *A shortest path problem.*

Let P^* be the shortest s–t path and denote its length by $L(P^*)$. Let P^{-S} be the shortest s–t path that does not use any edge belonging to an agent in $S \subset N$, and let $L(P^{-S})$ be its length. The substitutes condition (1) is the requirement

$$L(P^{-S}) - L(P^*) \geq \sum_{j \in S} [L(P^{-j}) - L(P^*)] \quad \forall S \subset N.$$

For this example $L(P^*) = 2$. The marginal product of agent A is $5 - 2 = 3$, which is the same as that of agent B. But for the coalition $S = \{1, 2\}$, the marginal product is also $5 - 2 = 3$. Since $3 < 3 + 3$, the substitutes condition does not hold. In this shortest path problem, agents A and B are complements; neither can be on an s–t path without the other.

6.1. Dual variables. While the substitutes condition need not hold in general in the shortest path environment, there are special cases where it does. One such example is in Schummer and Vohra (2001). When the substitutes condition holds, one can ask if there is a linear programming formulation whose dual variables give rise to agents' marginal products.[9] Here, we describe one such formulation.

To be consistent with the typical notation for this problem, for the remainder of this section we let A denote the set of agents. We assume that any s–t path uses at most one edge belonging to a particular agent $r \in A$.

[9] Hershberger and Suri (2001) provide an algorithm for finding Vickrey payments when valuations are given, even when the substitutes condition does not hold.

To formulate the shortest path problem as a linear program, let N be the node-arc incidence matrix of the graph; the column of N corresponding with the edge from vertices i to j contains a -1 in the row corresponding to i, a 1 in the row for j, and a 0 in the other rows. Let $x_e = 1$ if edge e is used, and zero otherwise. Then the problem is

$$\begin{aligned} \min \sum_{e \in E} c_e x_e & \\ \text{s.t. } Nx = b & \\ x_e \geq 0 \quad & \forall e \in E \\ \sum_{e \in E^r} x_e \leq 1 \quad & \forall r \in A \end{aligned}$$

where E^r is the set of edges belonging to agent $r \in A$.

Here, b is a column vector defined as $b_s = -1$, $b_t = 1$, and $b_i = 0$ for all $j \in V \setminus \{s, t\}$. The last set of constraints is redundant given the assumption that any s–t path uses at most one edge belonging to a particular agent. While redundant, some of them are binding in an optimal solution. This degeneracy in the primal implies multiple optimal dual solutions. Let P^* be the shortest path (and without loss of generality, assume that it is unique).

We claim that among all optimal dual solutions, there is one where for each $r \in A$, the dual variable associated with the constraint

$$\sum_{e \in E^r} x_e \leq 1$$

is agent r's marginal product.

To demonstrate this, let $q \in A$ be an agent who owns an edge in the shortest s–t path. To determine q's marginal product, we need to compute the length of the shortest s–t path that does not use any edge in E^q. Call this path P^{-q}. Hence agent q's marginal product is $L(P^{-q}) - L(P^*)$.

Let

$$\begin{aligned} Z(\mu) = \min \sum_{e \in E} c_e x_e & \\ \text{s.t. } Nx = b & \\ x_e \geq 0 \quad & \forall e \in E \\ \sum_{e \in E^r} x_e \leq 1 \quad & \forall r \in A \setminus q \\ \sum_{e \in E^q} x_e \leq \mu & \end{aligned}$$

Now, $Z(t)$ is piecewise linear in μ. It is easy to see that $\mu = 1$ is a breakpoint. For $\mu > 1$, the dual variable must be zero.

For $\mu < 1$, consider $\mu = 1 - \theta$ for $0 < \theta < 1$. The optimal solution involves a possibly fractional flow of one unit from s to t. By path-flow decomposition we can write this flow as a sum of flows along s–t paths. It is then easy to see that $1-\theta$ units must flow along P^* (the shortest path for $\mu = 1$), and the remaining θ units can flow along path P^{-q}. The remaining θ units cannot flow on any other path that intersects E^q because this would violate the last constraint.

Hence,

$$Z(\mu) = Z(1) - \theta[\sum_{e\in P^*} c_e - \sum_{e\in P^{-q}} c_e] = Z(1) + \theta[L(P^{-q}) - L(P^*)].$$

So the change in objective function value is exactly θ times agent q's marginal product, demonstrating the claim.

To write down the dual to the shortest path formulation, we associate a variable ν_i with each row of $Nx = b$ and λ_r with each constraint $\sum_{e\in E^r} x_e \le 1$. The dual is

$$\begin{aligned} &\max \nu_t - \nu_s - \sum_{r\in A} \lambda_r \\ &\text{s.t. } \nu_j - \nu_i - \lambda_r \le c_{ij} \quad \forall (i,j) \in E^r \\ &\lambda_r \ge 0 \quad \forall r \in A \end{aligned}$$

We prove that when agents are substitutes, there is an optimal solution to the dual where $\lambda_r = L(P^{-r}) - L(P^*)$ for all $r \in A$. First, observe that for any choice of $\lambda \ge 0$ we can always find ν to satisfy the dual constraints. In fact, for each choice of λ, let $\nu(\lambda)$ be the optimal solution to:

$$\begin{aligned} &\max \nu_t - \nu_s \\ &\text{s.t. } \nu_j - \nu_i \le c_{ij} + \lambda_r \quad \forall (i,j) \in E^r \end{aligned}$$

From now on, fix $\lambda_r = L(P^{-r}) - L(P^*)$ for all $r \in A$.

For any s–t path P, let A_P denote the set of agents whose edges are traversed in the path P. Then, for any path P, we have, by summing up the dual constraints $\nu_j - \nu_i - \lambda_r \le c_{ij}$ for every edge on this path:

$$\nu_t(\lambda) - \nu_s(\lambda) \le L(P) + \sum_{r\in A_P} \lambda_r.$$

In fact, because the linear program for $\nu(\lambda)$ corresponds to the dual of a shortest path problem with costs $c'_{ij} = c_{ij} + \lambda_r$ for $(i,j) \in E^r$, we know that there is a shortest path realizing $\nu_t(\lambda) - \nu_s(\lambda)$. Hence, we conclude that

$$\nu_t(\lambda) - \nu_s(\lambda) = \min_P\{L(P) + \sum_{r\in A_P} \lambda_r\}.$$

Weak duality implies that

$$L(P^*) \ge \nu_t(\lambda) - \nu_s(\lambda) - \sum_{r\in A} \lambda_r = \nu_t(\lambda) - \nu_s(\lambda) - \sum_{r\in A_{P^*}} \lambda_r.$$

Thus

$$L(P^*) \geq \min_P \{L(P) + \sum_{r \in A_P} \lambda_r\} - \sum_{r \in A_{P^*}} \lambda_r.$$

Let $S = A \setminus A_{P^*}$.

$$\min_P \{L(P) + \sum_{r \in A_P} \lambda_r\} = L(P^{-S}) + \sum_{r \in A_{P-S}} \lambda_r.$$

Then

$$L(P^*) \geq L(P^{-S}) + \sum_{r \in A_{P-S}} \lambda_r - \sum_{r \in A_{P^*}} \lambda_r = L(P^{-S}) - \sum_{r \in S} \lambda_r \geq L(P^*)$$

where the last inequality follows from the substitutes condition. Hence, $(\nu(\lambda), \lambda)$ constitutes an optimal dual solution.

7. Homogeneous goods. We now turn to the simplest case of K identical indivisible units of a product. We will consider two cases. In Section 7.1, each bidder wants at most one unit. In Section 7.2, each bidder may want multiple units.

7.1. Unit demand. Let v_j denote bidder j's value for one unit, and suppose it to be integral. Given such valuations, we can find the efficient allocation of the K units by solving a linear program.

Let $x_j = 1$ if an object is assigned to bidder j, and 0 otherwise.

$$\begin{aligned} V(N) = \max & \sum_{j \in N} v_j x_j \\ \text{s.t.} & \sum_{j \in N} x_j \leq K \\ & 0 \leq x_j \leq 1 \quad \forall j \in N \end{aligned}$$

Notice that this LP has all integral extreme points, so the optimal solution/allocation is integral. This formulation is clearly a special case of (P4).

Without loss of generality, assume that values are in decreasing order so that $v_j \geq v_{j+1}$. The optimal solution is $x_j = 1$ for $1 \leq j \leq K$ and $x_j = 0$ for $j \geq K + 1$.

Let y be the dual variable associated with the first constraint above. The dual to this LP is:

$$\begin{aligned} \min & \sum_{j \in N} \mu_j + Ky \\ \text{s.t. } & y + \mu_j \geq v_j \quad \forall j \in N \\ & y, \mu_j \geq 0 \end{aligned}$$

Complementary slackness requires that $y + \mu_j = v_j$ for $j = 1, \dots, K$. Given this, it is easy to see that one set of optimal dual variables is $y = v_{K+1}$, $\mu_j = v_j - y$ for $1 \leq j \leq K$, and $\mu_j = 0$ for $j \geq K + 1$.

Now consider the following auction. Bidders submit bids. Compute the optimal solution to the primal problem above, and make the assignment implied by it. Charge each bidder who gets an object the value of y found above.

The value of y corresponds exactly to the payment prescribed by the Vickrey auction. To see why, observe that in this case, $V(N) = \sum_{i=1}^{K} v_i$. Furthermore, $V(N \setminus j) = \sum_{i=1}^{K} v_i - v_j + v_{K+1} = V(N) - v_j + v_{K+1}$ for $j \leq K$, and $V(N \setminus j) = V(N)$ otherwise. Hence,

$$V(N) - V(N \setminus j) = [v_j - v_{K+1}]^+ = [v_j - y]^+.$$

In the dual, we can interpret each binding constraint $y + \mu_j = v_j$ as saying that if bidder j gets an object, then her surplus is $\mu_j = v_j - y$. The particular dual solution we have chosen is the one that maximizes the total surplus to the bidders:

$$\begin{aligned} &\min y \\ &\text{s.t.} \sum_{j \in N} \mu_j + Ky = z_{\text{lp}} \\ &\quad y + \mu_j \geq v_j \quad \forall j \in N \\ &\quad y, \mu_j \geq 0 \end{aligned}$$

The ascending auction that generates the same outcome is well known, and it is an equilibrium for everyone to bid sincerely in each round of that auction. It will be instructive to show how it arises naturally from a primal-dual algorithm for the problem.

7.1.1. The primal-dual algorithm for this setting. Here we show how the ascending Vickrey auction arises naturally from a primal-dual algorithm (see Algorithm 1). Most of the difficulty in describing the algorithm stems from the possibility of ties, i.e., different agents having identical valuations.

Without loss of generality, we assume that each bidder has a strictly positive value for the good. The dual problem is:

$$\begin{aligned} &\min \sum_{j \in N} \mu_j + Ky \\ &\text{s.t. } y + \mu_j \geq v_j \quad \forall j \in N \\ &\quad y, \mu_j \geq 0 \end{aligned}$$

Given a value $\hat{y}$ (initially chosen to equal 0) for y, a feasible dual solution can be constructed by setting $\hat{\mu}_j = [v_j - \hat{y}]^+$. Let $I = \{j : \hat{\mu}_j = 0\}$

and $J = \{j : \hat{\mu}_j + \hat{y} = v_j\}$. The elements of J correspond to bidders with nonnegative surplus. Note that $J \cup I = N$.

Bidders in $J \setminus I$ have strictly positive surplus, i.e., they strictly want to buy an object at the current price $\hat{y}$. Bidders in $J \cap I$ are indifferent between buying at the current price $\hat{y}$ (with zero surplus) and not buying. Bidders in $I \setminus J$ value the items at less than $\hat{y}$; they strictly prefer not to buy. In the auction environment, we would not know the sets I and J, and would have to rely on the reports of the bidders.

If this dual solution were in fact optimal, we know, by complementary slackness, that for every primal optimal solution $\bar{x}$,

1. $j \in I \Rightarrow \bar{x}_j \leq 1$,
2. $j \notin I \Rightarrow \bar{x}_j = 1$,
3. $j \in J \Rightarrow \bar{x}_j \geq 0$,
4. $j \notin J \Rightarrow \bar{x}_j = 0$ and
5. $y > 0 \Rightarrow \sum_{j \in N} \bar{x}_j = K$.

A primal solution satisfying these conditions exists if and only if both $|N \setminus I| \leq K$ and $|J| \geq K$. The restricted primal simply formulates, as a linear program, the problem of deciding if these inequalities holds. To ensure that the restricted primal is feasible, we introduce artificial variables x_j^a. The restricted primal **(RP)** is

$$\max -x_0^a - \sum_{j \in N \setminus I} x_j^a$$

$$\begin{aligned}
\text{s.t. } & \sum_{j \in N} \bar{x}_j + x_0^a = K \\
& \bar{x}_j + x_j^a = 1 \quad \forall j \in N \setminus I \\
& 0 \leq \bar{x}_j \quad \forall j \in N \setminus I \\
& \bar{x}_j = 0 \quad \forall j \in N \setminus J \\
& \bar{x}_j \leq 1 \quad \forall j \in I \cap J \\
& 0 \leq \bar{x}_j \quad \forall j \in I \cap J \\
& 0 \leq x_j^a \quad \forall j \in N \cup \{0\}
\end{aligned}$$

For $\hat{y} = 0$ the objective function would be $\max -\sum_{j \in N \setminus I} x_j^a - \sum_{j \in N \setminus J} x_j^a$.

If both $|N \setminus I| \leq K$ and $|J| \geq K$, then the optimal objective function value of (RP) is zero and we have found a primal solution that satisfies all complementary slackness conditions.

We show that the case $|J| < K$ (where there is too little demand) can be ignored, because the auction will end before it occurs. The first time (round) at which $|J| < K$ occurs, there must have been "overdemand" in the previous round, i.e., $K < |N \setminus I'|$ (assuming non-negative values v_j, where the prime signifies "previous round") otherwise the auction would have ended. Since $J' \cup I' = N$, this implies $|J' \setminus I'| > |J|$, which is a contradiction, since, from one round to the next, we must have $J' \setminus I' \subset J$.

In other words, from one round to the next, an index $j \in N$ can move from $J \setminus I$ to $I \cap J$ or from $I \cap J$ to $I \setminus J$; moving from $J \setminus I$ to $I \setminus J$ requires at least two rounds.

If $K - |N \setminus I| < 0$, then the optimal objective function value is $K - |N \setminus I| < 0$, so there is no primal solution that is complementary to the current dual solution. To determine how to update the dual variables we consider the dual to the restricted primal **(DRP)**:

$$\begin{aligned}
\min \sum_{j \in J} \mu_j + Ky& \\
\text{s.t. } y + \mu_j \geq 0 & \quad \forall j \in J \\
y \gtrless 0& \\
\mu_j \gtrless 0 & \quad \forall j \in (N \setminus I) \cup (N \setminus J) \\
\mu_j = 0 & \quad \forall j \in N \setminus J \\
\mu_j \geq -1 & \quad \forall j \in N \setminus I \\
\mu_j \geq 0 & \quad \forall j \in I \cap J \\
y \geq -1 \, (\text{or } y \geq 0 \text{ if } \hat{y} = 0)&
\end{aligned}$$

(The symbol $\gtrless$ is used to mean that the variable is unrestricted.) This problem can be simplified to

$$\begin{aligned}
\min \sum_{j \in J} \mu_j + Ky& \\
\text{s.t. } y + \mu_j \geq 0 & \quad \forall j \in J \\
\mu_j \geq -1 & \quad \forall j \in N \setminus I \\
\mu_j \geq 0 & \quad \forall j \in I \cap J \\
y \geq -1 \, (\text{or } y \geq 0 \text{ if } \hat{y} = 0)&
\end{aligned}$$

Denote the optimum to this problem with $(\bar{y}, \bar{\mu}^T)^T$. It is obvious, because $K < |N \setminus I| \leq |J|$, that $\bar{y} = 1$, $\bar{\mu}_j = -1$ for $j \in N \setminus I$ is the optimum (since it is feasible and its value of $K - |N \setminus I|$ matches the value of the (RP)). This corresponds to raising the price by one, decreasing the surplus of bidders with strictly positive surplus by one, and removing the bidders in $I \cap J$ from the auction.

We need to update the dual prices to: $(\hat{y}, \hat{\mu})^T \leftarrow (\hat{y}, \hat{\mu})^T + \theta(\bar{y}, \bar{\mu})^T$ for suitable $\theta > 0$. Notice that, as the optimum of (DRP) is strictly negative, this corresponds to raising the price by θ and forcing the bidders with strictly positive surplus to decrease their surplus by θ.

Clearly, θ has to be chosen to maintain feasibility of $(\hat{y}, \hat{\mu})^T + \theta(\bar{y}, \bar{\mu})^T$ in the dual. For $j \in J$ the constraints $y + \mu_j \geq v_j$ are certainly fulfilled.

For $j \in N \setminus J$ those constraints require $\hat{y} + \theta\bar{y} + \hat{\mu}_j + \theta\bar{\mu}_j \geq v_j$. If $\bar{y} + \bar{\mu}_j \geq 0$, the constraint is fulfilled for all $\theta > 0$. The case that $\bar{y} + \bar{\mu}_j < 0$ is by our choice of the dual solution impossible.

Similarly, we have to ensure the nonnegativity of the μ_j's. Again, this is a non-issue for $i \in I$ (because they are unchanged). For $i \in N \setminus I$ the condition is $\hat{\mu}_j + \theta\bar{\mu}_j \geq 0$ and this reduces to $\theta \leq -\hat{\mu}_j/\bar{\mu}_j$. So we could choose (if we know the bidder's μ_j's) a step size by

$$\theta = \min_{j \in N \setminus I} \frac{-\hat{\mu}_j}{\bar{\mu}_j}$$

By examination, we see that (for integral v_j), $\theta \geq 1$ unless the procedure is over.

7.2. Homogeneous goods, multi-unit demand. As before, there are K indivisible units to be auctioned off. In this case, though, bidders may want more than one object. Let v_i^t be the value that bidder i assigns to their t^{th} unit (conditional on having already received $t-1$ units). We assume that bidders have diminishing marginal values for units: $v_i^t \geq v_i^{t+1}$ for all $i \in N$ and $1 \leq t \leq K$.

7.2.1. A natural formulation. A natural formulation for the problem defines $x_{i,t} = 1$ if bidder i receives a tth object.

$$\begin{aligned}
\max \sum_{i \in N} \sum_{t=1}^{K} v_i^t x_{i,t} & \\
\text{s.t. } \sum_{i \in N} \sum_{t=1}^{K} x_{i,t} \leq K & \\
x_{i,t+1} \leq x_{i,t} \quad & \forall i \in N, \forall t < K \\
x_{i,1} \leq 1 \quad & \forall i \in N \\
x_{i,K} \geq 0 \quad & \forall i \in N
\end{aligned}$$

The dual variable on $x_{i,1}$ gives the surplus of agent i.

To the resource-constraint we associate a dual variable y, to each inequality $x_{i,t+1} - x_{i,t} \leq 0$ for $t < K$ a variable $\mu_{i,t+1}$, and with each inequality $x_{i,t} \leq 1$ a variable $\lambda_{i,t}$.

Now we can write down the dual formulation.

$$\begin{aligned}
\min Ky + \sum_{i \in N} \sum_{t=1}^{K} \lambda_{i,t} & \\
\text{s.t. } \lambda_{i,1} - \mu_{i,2} + y \geq v_i^1 \quad & \forall i \in N \\
\lambda_{i,t} - \mu_{i,t+1} + \mu_{i,t} + y \geq v_i^t \quad & \forall i \in N, \forall t < K \\
\lambda_{i,K} + \mu_{i,K} + y \geq v_i^K \quad & \forall i \in N \\
y \geq 0 & \\
\lambda_{i,t} \geq 0 \quad & \forall i \in N, \forall t \leq K \\
\mu_{i,t} \geq 0 \quad & \forall i \in N, \forall t > 1
\end{aligned}$$

The primal-dual algorithm for this formulation does not, however, implement the Vickrey outcome. In fact, it leads to the "uniform price" auction, in which all winning bidders pay the same price per unit: the highest losing bid.

7.2.2. Ausubel's ascending auction. For the case of homogeneous goods and multi-unit demand, Ausubel (1997) was the first to propose an ascending auction to implement the Vickrey outcome. Bikhchandani and Ostroy (2000b) were the first to show that Ausubel's auction was an instance of the primal-dual algorithm for formulation (P3). We have mentioned elsewhere in this paper that Ausubel's auction can be interpreted as a primal-dual algorithm for a matroid optimization problem. Here we outline how Ausubel's auction can be derived from a primal-dual algorithm for the longest path problem in an acyclic network. To do this, we refine a transformation from knapsack problems to longest paths problems; see Ahuja et al. (1997).

Let $V_r^t \equiv \sum_{\tau=1}^t v_r^\tau$ be agent r's total value for obtaining t units. We will formulate the problem of finding the efficient allocation as a longest path problem in an acyclic directed graph G.

Each vertex in G is identified by an ordered pair (r, s) of integers. A vertex (r, s) represents the state in which bidders 1 through $r - 1$ have together acquired s units. In addition, vertex $(1, 0)$ is the source and vertex $(N + 1, K)$ is the sink. Therefore the set of vertices is

$$\{(1,0)\} \cup \left(\bigcup_{r=2}^{N} \bigcup_{s=0}^{K} \{(r,s)\} \right) \cup \{(N+1,K)\}.$$

Edge $((r, s), (r + 1, s + t))$ belongs to bidder r when $t > 0$, and belongs to the auctioneer for $t = 0$.

For each vertex (r, s), there is an edge directed to each vertex $(r + 1, s + t)$ such that $0 \leq t \leq K - s$. Such an edge corresponds to bidder r obtaining t units. The length of that edge is V_r^t (where $V_r^0 \equiv 0$).

Each source-sink path in G corresponds to a feasible assignment of K objects among the bidders. Since G is acyclic, the longest path problem is well defined on it. Also each path from the source to the sink uses at most one edge belonging to any given agent. Since agents are substitutes in this environment, it follows from the shortest path section of this paper that the dual variables that correspond to the constraint that a bidder can own only one edge on a path is his marginal product (in a suitably chosen dual solution).

The problem of finding the longest path can be formulated as

$$\max \sum_{r=1}^{N-1} \sum_{t=0}^{K} \sum_{s=0}^{K-t} V_r^t x_{(r,s),(r+1,s+t)} + \sum_{s=0}^{K} V_N^{K-s} x_{(N,s),(N+1,K)}$$

$$\text{s.t.} \sum_{s=0}^{K} x_{(N,s),(N+1,K)} = 1$$

$$\sum_{t=0}^{s} x_{(r-1,s-t),(r,s)} - \sum_{t=0}^{K-s} x_{(r,s),(r+1,s+t)} = 0 \quad \forall s, \forall r > 2$$

$$x_{(1,0),(2,s)} - \sum_{t=0}^{K-s} x_{(2,s),(3,s+t)} = 0 \quad \forall s$$

$$\sum_{s=0}^{K} \sum_{t=1}^{K-s} x_{(r,s),(r+1,s+t)} \leq 1 \quad \forall 1 < r \leq N$$

$$\sum_{t=1}^{K} x_{(1,0),(2,t)} \leq 1$$

$$x_e \geq 0 \quad \forall e.$$

The first constraint requires that exactly one edge is chosen to enter the sink $(N+1, K)$. The second class of constraints requires that in each intermediate node, the number of entering arcs equals the number of leaving arcs; the third constraint requires the same for vertices whose incident arcs come from the source. The fourth class limits—to at most one—the number of edges giving bidder $r > 1$ a positive number of units, and the fifth is the same constraint for bidder 1. These last constraints are redundant.

To write down the dual, we associate with each constraint of type

- $\sum_{s=0}^{K} x_{(N,s),(N+1,K)} = 1$, an unrestricted variable $\nu_{N+1,K}$,
- $\sum_{t=0}^{s} x_{(r-1,s-t),(r,s)} - \sum_{t=0}^{K-s} x_{(r,s),(r+1,s+t)} = 0$, $\forall s$, $\forall r > 2$, an unrestricted variable $\nu_{r,s}$,
- $x_{(1,0),(2,s)} - \sum_{t=0}^{K-s} x_{(2,s),(3,s+t)} = 0$, $\forall s$, an unrestricted variable $\nu_{2,s}$,
- $\sum_{s=0}^{K} \sum_{t=1}^{K-s} x_{(r,s),(r+1,s+t)} \leq 1$, $\forall 1 < r \leq N$, a variable $\lambda_r \geq 0$,
- $\sum_{t=1}^{K} x_{(1,0),(2,t)} \leq 1$, a variable $\lambda_1 \geq 0$.

The dual is

$$\min \nu_{N+1,K} + \sum_{r=1}^{N} \lambda_r$$

$$\text{s.t. } \nu_{r+1,s+t} - \nu_{r,s} + \lambda_r \geq V_r^t \quad \forall t \geq 1, \forall (r,s)$$

$$\nu_{r+1,s} - \nu_{r,s} \geq 0 \quad \forall (r,s)$$

$$\lambda_r \geq 0 \quad \forall r \in N$$

$$\nu_{r,s} \gtrless 0 \quad \forall (r,s)$$

$$\nu_{1,0} = 0$$

A feature of the construction to be noted is that a bidder $r > 1$ owns many different edges that correspond to consuming t units of the good. Bidder 1, however, has exactly one edge that corresponds to consuming t units. We want to equally price all edges that correspond to bidder $r > 1$ consumes t units. To do this, we later ensure that for $r > 1$, $\nu_{r+1,s+t} - \nu_{r,s} = \nu_{r+1,s+t+1} - \nu_{r,s+1}$.

For an initial feasible solution, set $\hat{\nu} = 0$, corresponding to initial prices on all items of zero. Let $\hat{\lambda}_r = \max_{s,t}[V_r^t - \hat{\nu}_{r+1,s+t} + \hat{\nu}_{r,s}]^+$. Let $I = \{r \colon \hat{\lambda}_r = 0\}$ be the set of agents without positive surplus, and let bidder r's "demand set" be denoted

$$J^r = \{((r,s),(r+1,s+t)) \colon t \geq 1 \wedge \nu_{r+1,s+t} - \nu_{r,s} + \lambda_r = V_r^t\}.$$

This is the set of edges that maximize the surplus of bidder r. Denote the set of edges that bidder r is "willing" to consume as

$$L^r = \{((r,s),(r+1,s+t)) \colon t \geq 1 \wedge \nu_{r+1,s+t} - \nu_{r,s} \leq V_r^t\}.$$

These edges leave their owner with non-negative surplus.

If $L^r = \emptyset$ set $\alpha_r = 0$. Otherwise set $\alpha_r = \max\{t : ((r,s),(r+1,s+t)) \in J^r\}$. The value α_r is the greatest number of items that bidder r wants to buy at current prices. Similarly, set $\beta_r = 0$ if $L^r = \emptyset$ and otherwise $\beta_r = \min\{t : ((r,s),(r+1,s+t)) \in J^r\}$. The value β_r is the smallest number of items that bidder r wants to buy at current prices. If $\sum_r \beta_r \leq K \leq \sum_r \alpha_r$ we have the optimal solution and we are done.

Using the current dual values, determine for each bidder r the number $\gamma_r = [K - \sum_{p \neq r} \beta_p]^+$. Notice that γ_r is the number of units that bidder r has to buy so that all items are sold. Ausubel calls this the number of items *clinched* by bidder r. Let $\gamma = K - \sum_{r=1}^{n} \gamma_r$ be the total number of nonclinched items.

Let E^r denote the set of edges owned by agent r and E^0 be the edges that the auctioneer holds. If the current dual solution is optimal, we know, by complementary slackness, for every primal optimal solution $\bar{x}$,

1. $r \in I \Rightarrow \sum_{e \in E^r} \bar{x}_e \leq 1$,
2. $r \notin I \Rightarrow \sum_{e \in E^r} \bar{x}_e = 1$,
3. $e \in J^r \Rightarrow \bar{x}_e \geq 0$, and
4. $e \in E^r \setminus J^r \Rightarrow \bar{x}_e = 0$.

The corresponding restricted primal **(RP)** will be:

$$\max -(y^+_{N+1,K} + y^-_{N+1,K}) - \gamma \sum_{r=1}^{N} \sum_{s=0}^{K} (y^+_{rs} + y^-_{rs})$$

$$\text{s.t.} \sum_{s=0}^{K} x_{(N,s),(N+1,K)} + (y^+_{N+1,K} - y^-_{N+1,K}) = 1$$

$$\sum_{t=0}^{s} x_{(r-1,s-t),(r,s)} - \sum_{t=0}^{K-s} x_{(r,s),(r+1,s+t)} + (y^+_{rs} - y^-_{rs}) = 0 \quad \forall s, \forall r > 2$$

$$x_{(1,0),(2,s)} - \sum_{t=0}^{K-s} x_{2,s,3,s+t} + (y^+_{1s} - y^-_{1s}) = 0 \quad \forall s$$

$$x_e \geq 0 \quad \forall e \in E$$

$$\sum_{e \in E^r} x_e \leq 1 \quad \forall r \in I$$

$$\sum_{e \in E^r} x_e = 1 \quad \forall r \notin I$$

$$x_e \geq 0 \quad \forall e \in J^r$$

$$x_e = 0 \quad \forall e \in E^r \setminus J^r$$

$$y \geq 0$$

Denote by $G(J)$ a graph with the same set of vertices as G but only those edges that belong to J. If an arc $(r, s), (r + 1, s + t)$ is present in $G(J)$ this means that under current prices, it is a surplus-maximizing arc for agent r. That is at current prices, agent r maximizes his surplus by buying t units.

Note that (RP) has value zero if and only if $G(J)$ contains a source-sink path such that each agent from $N \setminus I$ owns one edge on this path. If there is no such path from source to sink in $G(J)$, then we have two possible causes. First, it could be that not even $G(L) \supseteq G(J)$ itself contains a source-sink path. Fortunately, we can avoid this case: In the first iteration, prices are 0 and $G(L)$ is connected. It suffices to keep $G(L)$ connected in subsequent iterations, and we do that below.

Second, it could be that $G(L)$ contains a path, but no such path contains, for each bidder $r \in N \setminus I$, an edge in J^r. In this case, agents demand too much at current prices. To deal with this, we raise item prices, to make large quantities (that is, long arcs) less attractive to agents so that they demand smaller quantities, which corresponds to demanding arcs $(r, s), (r + 1, s + t)$ with smaller t's.

Consider now **(DRP)**, the dual of the restricted primal.

$$\min \nu_{N+1,K} + \sum_{r=1}^{N} \lambda_r$$

$$\text{s.t. } \nu_{r+1,s+t} - \nu_{r,s} + \lambda_r \geq 0 \quad \forall((r,s),(r+1,s+t)) \in J^r$$

$$\nu_{r+1,s} - \nu_{r,s} \geq 0 \quad \forall(r,s)$$

$$\lambda_r \geq 0 \quad \forall r \in I$$

$$\lambda_r \gtreqless 0 \quad \forall r \notin I$$

$$\nu_{1,0} = 0$$

$$\nu_{r,s} \geq -\gamma \quad \forall(r,s)$$

$$\nu_{r,s} \leq \gamma \quad \forall(r,s)$$

We need to find a feasible solution $(\bar{\nu}, \bar{\lambda})$ that results in a negative objective function value. As long as no bidder has clinched anything, we uniformly raise item prices by $\theta > 0$. As soon as bidder r clinches some items, the price on these items is fixed for bidder r. Choosing the direction in which the dual variables are to be adjusted takes care of this: $\bar{\nu}_{r,s} = [s - \sum_{q=1}^{r-1} \gamma_q]^+$. By computing $\bar{\lambda}$ accordingly, one can show that we have a negative objective value.

Consider (in the case of no truncation) the difference $\bar{\nu}_{r+1,s+t} - \bar{\nu}_{r,s} = t - \gamma_r$. We raise the price for bundles that bidder r obtains only by the number of non-clinched items in it. This is analogous to Ausubel's ascending auction.

Choosing this dual direction permits us to avoid the uniform prices and steer towards the efficient outcome. From the substitutes property, we know that the final values of the dual variables will give to each agent a payoff that equals his marginal product.

8. Summary. This paper surveys the connections between sealed bid Vickrey auctions and duality in linear programming. By example, we have shown how this relation can be exploited to produce iterative auctions that implement the Vickrey outcome in various scenarios. The approach can be summarized by the following steps.

1. Verify that the *agents are substitutes* condition holds.
2. Formulate a linear program where appropriate dual variables correspond to the marginal products of the agents. Under the substitutes condition, an optimal dual solution exists in which, *simultaneously*, each appropriate variable takes on the value of the corresponding agent's marginal product.
3. Construct a primal-dual algorithm for the linear program. The algorithm must choose an improving direction in the dual that will cause the algorithm to terminate in the dual solution described above. This is the one that, among all optimal dual solutions, is the one that maximizes the combined surplus of the bidders.

If the substitutes condition does not hold, we believe it is unlikely that an iterative auction (in which bidding sincerely is an equilibrium) yielding the Vickrey outcome exists. This is not a formal statement, since the term "iterative auction" requires a precise definition. So, an interesting line of inquiry would be to formalize this last intuition in a precise way.

REFERENCES

[1] Ahuja, R.K., T.L. Magnanti, and J.B. Orlin, Network Flows, Prentice Hall, New Jersey, 1993.

[2] Ausubel, L. (1997): "An Efficient Ascending-Bid Auction for Multiple Objects," http://www.ausubel.com/auction-papers/97wp-efficient-ascending-auction.pdf. Working Paper 97-06, Department of Economics, University of Maryland.

[3] Bertsimas, D. and C.-P. Teo (1998): "From Valid Inequalities to Heuristics: A Unified View of Primal-dual Approximation Algorithms in Covering Problems," *Operations Research*, 46, 503–514.

[4] Bikhchandani, S. and J. Ostroy (2000a): "The Package Assignment Model," mimeo, U.C.L.A.

[5] ——— (2000b): "Ascending Price Vickrey Auctions," mimeo, U.C.L.A.

[6] Clarke, E. (1971): "Multipart Pricing of Public Goods," *Public Choice*, 8, 19–33.

[7] Crawford, V.P. and E. Knoer (1981): "Job matching with Heterogeneous Firms and Workers," *Econometrica*, 49, 437–450.

[8] Demange, G., D. Gale, and M. Sotomayor (1986): "Multi-item auctions," *Journal of Political Economy*, 94, 863–72.

[9] Edmonds, J. and D.R. Fulkerson (1965): "Transversal and Matroid Partition," *Journal of Research of the National Bureau of Standards*, B69, 147–153.

[10] Groves, T. (1973): "Incentives in Teams," *Econometrica*, 41, 617–631.

[11] Gul, F. and E. Stacchetti (1999): "Walrasian Equilibrium with Gross Substitutes," *Journal of Economic Theory*, 87, 95–124.

[12] ——— (2000): "The English Auction with Differentiated Commodities," *Journal of Economic Theory*, 92, 66–95.

[13] Hoffman, A.J. and H. Markowitz (1963): "A Note on Shortest Path, Assignment and Transportation Problems," *Naval Research Logistics*, 10, 375–379.

[14] Hershberger, J. and S. Suri (2001): "Vickrey Pricing in Network Routing: Fast Payment Computation," mimeo, U.C. Santa Barbara.

[15] Kelso, A.S. and V.P. Crawford (1982): "Job matching, coalition formation, and gross substitutes," *Econometrica*, 50, 1483–1504.

[16] Krishna, V. and M. Perry (1998): "Efficient Mechanism Design," http://econ.la.psu.edu/~vkrishna/Papers/vcg19.pdf, mimeo, The Pennsylvania State University.

[17] Leonard, H. (1983): "Elicitation of Honest Preferences for the Assignment of Individuals to Positions," *Journal of Political Economy*, 91, 1–36.

[18] Moulin, H. (1986): "Characterizations of the Pivotal Mechanism," *Journal of Public Economics*, 31, 53–78.

[19] Papadimitriou, C.H. and K. Steiglitz (1982): "Combinatorial Optimization: Algorithms and Complexity," Prentice–Hall.

[20] Parkes, D.C. and L.H. Ungar (2000): http://www.cis.upenn.edu/ dparkes/pubs/ibundle00.ps"Iterative Combinatorial Auctions: Theory and Practice," in Proc. 17th National Conference on Artificial Intelligence (AAAI-00), 74–81.

[21] Schummer, J. and R.V. Vohra (2001): "Auctions for Procuring Options," http://www.kellogg.nwu.edu/faculty/schummer/ftp/research/optauc.pdf, mimeo, Northwestern University.

[22] Vickrey, W. (1961) "Counterspeculation, auctions, and competitive sealed tenders," *Journal of Finance*, 16, 8–37

[23] Williams, S.R. (1999): "A Characterization of Efficient, Bayesian Incentive Compatible Mechanisms," *Economic Theory*, 14, 155–180.

AUCTIONING TELECOMMUNICATIONS BANDWIDTH WITH GUARANTEED QUALITY OF SERVICE*

G. ANANDALINGAM† AND N.J. KEON‡

Abstract. In this paper, we present an auction mechanism for obtaining prices for different classes of telecommunications service with guaranteed quality of service (QoS). We model the problem as a nonlinear integer program, and use optimality conditions to direct our search for both centralized optimal results, and a mechanism for setting up an auction. Further, we examine multiple services, each with different QoS requirements including delay and packet loss, resource requirements including buffer and bandwidth needs, and origin-destination addresses. At the end, our methodology yields simple rules/tables for obtaining prices and resource allocations for these multiple "commodities."

1. Introduction. In this paper, we present a model for estimating prices for different classes of telecommunications service with guaranteed quality of service (QoS). In previous work we presented a model for load balancing in congested networks using price discounts [5] and also examined various ways of using iterative procedures for pricing networks with guaranteed QoS [6]. This paper extends our research in [6] and presents the results of using an auctioning method for pricing and resource allocation for multiple services in a telecommunications network with quality of service guarantees.

Pricing of telecommunications services has been a topic of research in the recent past. We will review the literature very briefly. Much of the work on pricing for packet-switched networks offering best-effort service has focused on so-called incentive compatible pricing [12] [13]. It has also been shown through simulations that priority pricing improved network performance when there was either a single class or multiple service classes [1]. It has also been shown that by offering a number of routes, with a corresponding set of *relative* discount rates, that a network can elicit users to select routes for data traffic according the desired operating point of the network provider [8]. Other work examine the methods for marking individual packets at congested resources so that shadow prices could be estimated at individual resources in a network [3] [7]. A simple dynamic pricing mechanism was proposed in [14].

There has been much less work published concerning pricing for networks offering QoS guarantees. Typically a network with guaranteed QoS

*This research was partially funded by a grant from the National Science Foundation NCR-9612781. Experiments were conducted in a course at the University of Maryland; the participation of students was invaluable for the paper. Previous versions of this paper has been presented in seminars at George Mason University, University of Maryland, and at the 2000 INFORMS meeting in Philadelphia. Comments from Mike Ball, Terry Friesz, and Steve Gabriel are acknowledged without implication.

†R.H. Smith School of Business, University of Maryland, College Park, MD 20742.

‡Edwin L. Cox School of Business, Southern Methodist University, Dallas, TX 75275.

must employ a call admission policy in order to satisfy the guarantees to users. In an alternative approach to call admission, it has been suggested that users guarantee their own QoS by purchasing the required bandwidth and buffer resources for their desired QoS directly from the network [10] [11]. A negotiation based framework for allocating network resources, using effective bandwidth as a base for pricing was proposed in [4]. In [10], Low and Varaiya have proposed another approach offering network resources such as bandwidth and buffer space directly to users as part of a bidding process. De Veciana and Baldick [16] study an analogous method based on announced prices.

The pricing decision for a single link point-to-point integrated services network over ATM was formulated as a constrained optimal control problem and a three-stage solution procedure was developed to calculate a price schedule in [17]. In [15], an analysis of a market-based methodology is offered as evidence that pricing schemes can offer efficient resource allocations in connection-oriented networks offering QoS guarantees. These models feature a conventional view of pricing each connection and announcing these prices to the public.

The results in this paper are very different in both scope and methodology in that we model the problem as an expected revenue maximizing model for the network provider. We use optimality conditions for the resulting nonlinear integer program to direct our search for both centralized prices and allocations, and for setting up an auction mechanism. Further, we examine multiple services, each with different service requirements such as buffer and bandwidth needs, and origin-destination addresses. Our methodology yields simple rules/tables for obtaining prices and resource allocations for these multiple "commodities."

1.1. Organization of the paper. The paper is organized as follows: In Section 2, we provide an overview of modeling features. In Section 3, we present the optimization problem from the network operator's point-of-view and discuss the optimality conditions. In Section 4, we present the network auction procedure. Section 5 discusses results from applying the schemes outlined in this paper for a small example. We end the paper with concluding remarks in Section 6.

2. Overview of some modeling features.

2.1. Service class, demand and pricing. In this section, we will discuss some of the main features of our model. These features, which will be part of the constraint set in the optimization problem apply to all types of network topologies. However, for ease of exposition, let us consider a ring type network in Figure 1. The switch processes different types of telecommunications traffic that either originates at the switch, or else arrives at the switch from other originating nodes. Traffic outputs from the switch either go along an outgoing bandwidth pipe, or directly to a user connected to that switch. A service *class* is denoted by the triple, (i,j,k) with service type i, origin, j, and destination, k. Each service type i has traffic characteristics defined by *average* bandwidth r_i and *peak* bandwidth R_i. The network operator provides resources (buffer and bandwidth) to complete the request for a service class; i.e. to allow the flow of a particular type of traffic between an origin-destination pair with guaranteed quality-of-service.

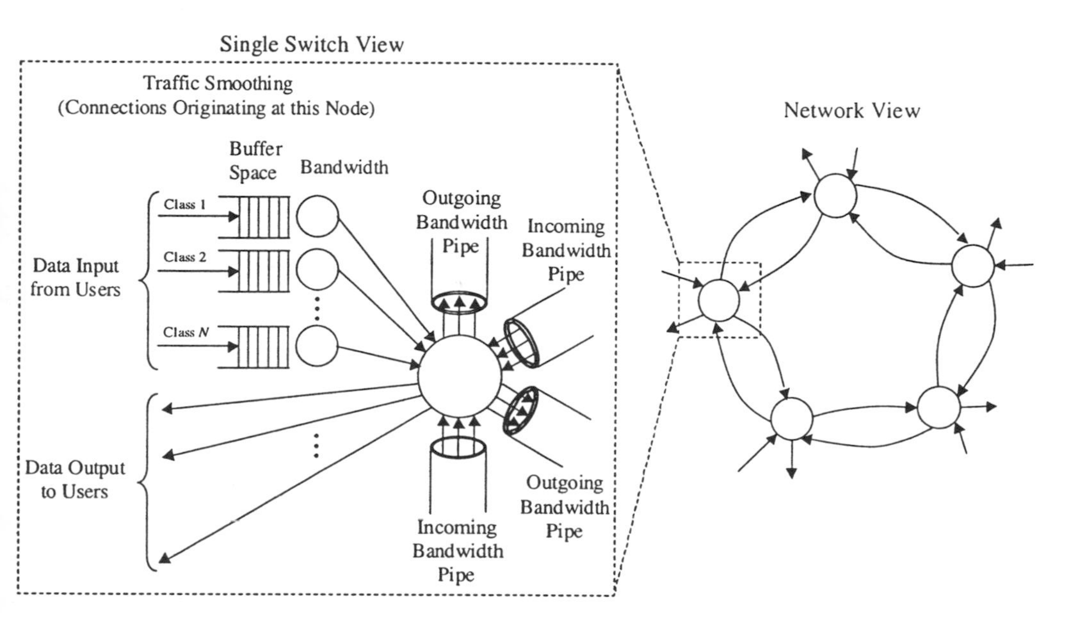

FIG. 1. *Network (ring) service model at an individual switch.*

The stochastic quantity demanded by users, λ_{ijk}, is taken to be the arrival rate of *connection* requests for service i, with the connection originating at switch j and terminating at switch k. This quantity is determined by a demand function $f_{ijk}(p_{ijk})$ where p_{ijk} is the price charged per unit of time the connection is open. We denote the elasticity of demand for service (i,j,k) by ε_{ijk} in the demand function below:

$$\lambda_{ijk} = a_{ijk} p_{ijk}^{-\varepsilon_{ijk}} \tag{1}$$

Note that the price charged p_{ijk} depends on the type of service i, and origin-destination pair (j,k), and can be determined centrally or through a decentralized process. In the centralized case, the arrival of connection requests given by (1) is a variable known with uncertainty. The network operator uses optimality conditions to obtain optimal prices and allocations. The decentralized process, described in this paper, involves using an auction process where bids are made by entering traffic (or the sources of this traffic) for resources to complete a type of service between an origin-destination pair. The bids involve both a willingness-to-pay (or price) and the number of connections required (and hence buffer/bandwidth) of a particular class of service. The network operator can either accept this bid and allocate the resources, or reject the bid and ask for a revision according to some rules.

2.2. Blocking traffic for quality-of-service. Although the network operator would maximize revenue by providing network access to all who want it, it would be difficult to guarantee quality-of-service due to limitation in buffer and bandwidth availability. Thus, some connection requests, especially those with low price bids and/or high resource requirements will have to be *blocked.* A centralized network operator would not know the connection arrival rate, and would do the analysis based on the assumption that for each service class, denoted by (i, j, k), the number of ongoing connections is an $M/GI/m/m$ queue where the arrival rate of connection requests, given by λ_{ijk} has inter-arrival times that are independently and identically distributed exponential random variables. The average holding time of a connection, T_{ijk}, is assumed known, and the holding times of individual connections can occur according to any general distribution, but are assumed to be independently and identically distributed for each connection. The traffic intensity of a class, ρ_{ijk}, is the product of arrival rate and average holding time, i.e., $\rho_{ijk} = \lambda_{ijk}T_{ijk}$. The capacity of any particular service class is $NCMAX_{ijk}$.

Using the well established properties of the $M/GI/m/m$ queuing model, we get that the probability of blocking a request within any service class, BP_{ijk}, is given by:

$$(2) \quad BP_{ijk} = P\,(NC_{ijk} = NCMAX_{ijk}) = \frac{\rho_{ijk}^{NCMAX_{ijk}}}{NCMAX_{ijk}!} \Bigg/ \sum_{n=0}^{NCMAX_{ijk}} \frac{\rho_{ijk}^{n}}{n!}$$

The expected number of connections is given by:

$$(3) \qquad E[NC_{ijk}] = \rho_{ijk}\,(1 - BP_{ijk})$$

If the number of open connections, NC_{ijk}, is equal to the maximum number permitted for each class, $NCMAX_{ijk}$, when a new user's request for service is received, the new connection will be blocked and lost to the system. Otherwise the request is admitted.

2.3. Limits on packet dropping for admitted traffic. In addition to connection (or call) blocking, one can drop packets and continue to maintain QoS for some classes. Some types of traffic, namely voice and video, can tolerate some packet loss; however, we need to set an upper limit on this. At the switch, packets from all connections of the same class

(i,j,k) share a FIFO queue of size B_{ijk}, the size of the input buffer which is a decision variable in our model. Note that this buffer can allow up to $NCMAX_{ijk}$ connections. Packet loss, $PLOSS_{ijk}$, occurs when there is buffer overflow.

There are a number of ways of limiting the packet loss for service classes that can tolerate it. In our model, we require that the minimum bandwidth allocation per offered connection be at least equal to the *equivalent capacity.* As we have discussed above, telecommunications traffic, whether voice, data or video, is described by two parameters, average bandwidth and peak bandwidth. The concept of "equivalent capacity" was derived in the telecommunications literature to present a single measure that best captures the "correct" combination of average and peak bandwidths. If bandwidth in excess of the equivalent capacity is assigned to a connection, the observed packet loss is less than $PLOSS_{ijk}$.

We use the results from [2] to calculate the equivalent capacity per connection, where all connections of a service class (i,j,k) share a single FIFO buffer. Consider a connection of a given service *type* i. The user is either not in the system (the "off" state) or sending data at a *peak* transmission rate, R_i, with an *average* rate given by r_i (the "on" state). Using the same model as in [2], we will assume that the on and off periods are exponentially distributed, with the average length of an on periods given by b_i. The traffic from all connections of the same type, i, is statistically identical, i.e., all connections classified as the same type have the same parameters r_i, R_i, and b_i. The equivalent capacity results from [2] are summarized below:

$$c(NCMAX_{ijk}, B_{ijk}, PLOSS_{ijk}) = \min\left(\hat{c}, \frac{\mu + \alpha'\sigma}{NCMAX_{ijk}}\right) \tag{4}$$

where,

$$\hat{c} = \frac{\alpha R_i - B_{ijk} + \sqrt{(\alpha R_i - B_{ijk})^2 + 4\alpha B_{ijk} r_i}}{2\alpha} \tag{5}$$

$$\alpha = \ln(1/PLOSS_{ijk})\, b_i \left(1 - \frac{r_i}{R_i}\right) \tag{6}$$

$$\mu = r_i NCMAX_{ijk} \tag{7}$$

$$\sigma = \sqrt{r_i (R_i - r_i)\, NCMAX_{ijk}} \tag{8}$$

$$\alpha' = \sqrt{-2\ln(PLOSS_{ijk}) - \ln(2\pi)} \tag{9}$$

The equivalent capacity per connection, (4), is calculated as the minimum of two distinct approximations; the first term is the effective bandwidth approximated by (5). The second term in (4) reflects multiplexing gains. This approximation is based on the stationary bit rate. To calculate this expression, we need the additional expression (6), as well as the mean of the aggregate bit rate, (7), the standard deviation of the aggregate bit rate, (8), and an approximate inversion of the normal distribution, (9). The calculations above separate equivalent capacity into regions dominated by either smoothing effects in the buffer or multiplexing gains.

As we shall see in the next section, the data traffic model and equivalent capacity expressions in (4)–(9) are sufficient for completing the formulation of our revenue maximizing model. It should be noted that our approach remains unchanged for *any other* derivation of equivalent capacity, which similarly to (4)–(9) satisfies the following general properties for bandwidth allocation:

$$r_i \leq c(NCMAX_{ijk}, B_{ijk}, PLOSS_{ijk}) \leq R_i \tag{10}$$

$$\frac{\partial c(NCMAX_{ijk}, B_{ijk}, PLOSS_{ijk})}{\partial B_{ijk}} \leq 0 \tag{11}$$

$$\frac{\partial c(NCMAX_{ijk}, B_{ijk}, PLOSS_{ijk})}{\partial NCMAX_{ijk}} \leq 0 \tag{12}$$

$$\frac{\partial c(NCMAX_{ijk}, B_{ijk}, PLOSS_{ijk})}{\partial PLOSS_{ijk}} \leq 0 \tag{13}$$

The first property, (10) states that the equivalent capacity must be greater than or equal to the mean rate of data traffic, r_i, and less than or equal to the peak rate of data traffic, R_i. The notion of effective bandwidth is captured by (11), with equivalent capacity decreasing in the amount of allocated buffer space, B_{ijk}. As the number of connections for which we allocate resources, $NCMAX_{ijk}$, increases, (12) reflects multiplexing gains. Finally, as we allow greater loss probabilities, $PLOSS_{ijk}$, the equivalent capacity decreases, as stated in (13). Note that equivalent capacity, as defined here, gives the bandwidth required to meet a single QoS criteria, namely loss probability, $PLOSS_{ijk}$. The inequalities admit special cases such as constant bit rate traffic, with no smoothing or multiplexing gains.

2.4. Limits on delay for admitted traffic. As we have noted, some traffic such as data can tolerate delay, and some, like voice, cannot. We also specify maximum delay, d_{ijk}, as a QoS parameter guaranteed to users. We ignore transmission delay and consider only delay in the buffer. Therefore we can set bounds for the potential delay a packet may experience quite easily:

$$d_{ijk} \leq \frac{B_{ijk}}{BW_{ijk}} \tag{14}$$

The maximum delay any packet may experience, given by (14), is simply the size of the allocated buffer space divided by the allocated bandwidth. The buffer is served on a first-in first-out (FIFO) basis.

3. Network operator's optimization problem.

3.1. The mathematical program. The network operator wants to maximize expected revenue subject to meeting all the QoS guarantees for the connections admitted to the network. If there was central control, the decision variables for the network operator are the price, p_{ijk}, for each service class (i,j,k), the volume of service (i.e. number of connections) offered, $NCMAX_{ijk}$. The amount of bandwidth, BW_{ijk}, reserved all along the path between j and k, and the amount of buffer space B_{ijk}, that should be reserved at the origin for each service class.

Recall that QoS is guaranteed by ensuring probability of loss, $PLOSS_{ijk}$, delay, d_{ijk}, and blocking with probability, BP_{ijk} are satisfied within certain limits. The optimization model from the standpoint of the network operator is:

(P)

$$\max_{p_{ijk},B_{ijk},BW_{ijk},NCMAX_{ijk}} \sum_{i=1}^{N}\sum_{j=1}^{NS}\sum_{k=1}^{NS} REV_{ijk}(p_{ijk},\lambda_{ijk}(p_{ijk}),BP_{ijk},T_{ijk}) \tag{15}$$

subject to,

$$BP_{ijk}\left(\lambda_{ijk}\left(p_{ijk}\right),T_{ijk},NCMAX_{ijk}\right) \le \overline{BP}_{ijk} \quad 1 \le i \le N, 1 \le j \le NS, 1 \le k \le NS \tag{16}$$

$$c\left(NCMAX_{ijk},B_{ijk},\overline{PLOSS}_{ijk}\right) \le \frac{BW_{ijk}}{NCMAX_{ijk}} \quad 1 \le i \le N, 1 \le j \le NS, 1 \le k \le NS \tag{17}$$

$$\frac{B_{ijk}}{BW_{ijk}} \le \bar{d}_{ijk} \quad 1 \le i \le N, 1 \le j \le NS, 1 \le k \le NS \tag{18}$$

$$\sum_{i=1}^{N}\sum_{j=1}^{NS}\sum_{k=1}^{NS} BW_{ijk} IVP_{ijkx} \le \overline{BW_x} \qquad 1 \le x \le NS \tag{19}$$

$$IVP_{ijkx} = \begin{cases} 1 & ,if\, x \in VP_{ijk} \\ 0 & ,otherwise \end{cases} \quad 1 \le i \le N, 1 \le j \le NS, 1 \le k \le NS, 1 \le x \le NS \tag{20}$$

$$\lambda_{ijk} \ge 0, \qquad p_{ijk} \ge 0, \qquad BW_{ijk} \ge 0, \qquad B_{ijk} \ge 0 \tag{21}$$

$$NCMAX_{ijk} \ge 0, \qquad \text{integer} \tag{22}$$

where,

REV_{ijk} = rate of revenue generation associated with service class (i, j, k)
c =equivalent capacity of a single connection
VP_{ijk} = {set of all x in the path for class i originating at j and terminating at $k, 1 \leq x \leq NS$}
$\overline{BP}_{ijk}$ = maximum blocking probability for a connection from service class (i, j, k)
$\overline{PLOSS}_{ijk}$ = maximum packet loss probability for a connection of service class (i, j, k)
$\bar{d}_{ijk}$ = maximum delay for a connection of service class (i, j, k)
$\overline{BW_j}$ = the capacity (bandwidth) at switch j.

The objective (15) is to maximize the average rate of revenue generation from ongoing connections, which is given by price times the *expected* number of connections. Constraint (16) ensures that the call-blocking probability for every class is below some prescribed limit. Constraint (17) ensures the probability of loss for each service class, $PLOSS_{ijk}$, is satisfied. The buffer delay for data traffic is constrained to be less than or equal to the limit given by the QoS guarantee, (18). Constraint (19) presents an upper limit to bandwidth available at each switch. There is an indicator function, (20), for every class (i,j,k), which indicates if any switch is included in the path. Constraints (21) are the usual non-negativity constraints, and (22) is the integrality constraint on the maximum number admitted to each class, $NCMAX_{ijk}$.The budget parameters, denoted by a bar overhead, are set outside the problem by the service provider.

3.2. Optimality properties for centralized decision making. The problem given in (P) is a nonlinear mixed integer problem. If the network operator acts as the central authority, there are a number of necessary and sufficient conditions that need to be satisfied. We will show that these conditions can be decomposed into those that apply to each service class independent of all other service classes. Based on this property, we can design an auctioning mechanism that allows for decentralized determination of prices and resource allocations.

We cannot prove that the problem given in P is convex. However, the following result will completely characterize the local optimal conditions:

THEOREM. *Given downward-sloping demand curves and plentiful buffer space at each originating switch, if an optimal solution to (P) does not coincide with marginal revenue equal to zero, i.e. the partial derivatives of (15) with respect to prices are all less than zero at optimality, then the optimal solution must satisfy the following properties:*

i)

$$BP_{ijk}\left(\lambda_{ijk}\left(p_{ijk}\right), T_{jk}, NCMAX_{ijk}\right) = \overline{BP}_{ijk} \quad 1 \leq i \leq N, 1 \leq j \leq NS, 1 \leq k \leq NS \tag{23}$$

ii)

$$\frac{BW_{ijk}}{NCMAX_{ijk}} = c_{ijk}\left(NCMAX_{ijk}, B_{ijk}, \overline{PLOSS}_{ijk}\right) \tag{24}$$
$$1 \le i \le N, 1 \le j \le NS, 1 \le k \le NS$$

iii)

$$\left(NCMAX_{xjk}+1\right)c_{xjk}\left(NCMAX_{xjk}+1, B_{xjk}, \overline{PLOSS}_{xjk}\right)$$
$$+\sum_{\substack{i=1\\ i\neq x}}^{N}\sum_{k=1}^{NS} NCMAX_{ijk}c_{ijk}\left(NCMAX_{ijk}, B_{ijk}, \overline{PLOSS}_{ijk}\right) > \overline{B}W_j \tag{25}$$
$$1 \le x \le N, 1 \le j \le NS$$

iv)

$$B_{ijk} = \bar{d}_{ijk}BW_{ijk} \qquad 1 \le i \le N, 1 \le j \le NS, 1 \le k \le NS \tag{26}$$

Proof. See Keon [6] for proof. □

The key assumption required for the conditions (23)–(26) to hold is that marginal revenue is less than zero, at optimality. If the marginal revenue for any service class were positive, a price increase would simultaneously reduce the use of resources and decrease revenue. On the other hand, with negative marginal revenue price decreases would increase revenue. Therefore, the assumption that marginal revenue is less than zero implies capacity (bandwidth) is completely assigned. In the case where marginal revenue equals zero at optimality, the problem is solved easily by setting the partial derivatives of the objective function with respect to prices (i.e. marginal revenue) equal to zero.

The optimality conditions given in the above Theorem shows that the call-blocking probability for all service classes is set to its greatest permitted value for all service classes in (23). Constraint (24) states that the bandwidth assigned to each service class, for the maximum number of connections admitted, will be equal to the corresponding equivalent capacity. The third property, (25), states that allocation of the equivalent capacity for any additional connections would violate feasibility, subject to the integrality of the maximum permitted connections. This means that no further increases in bandwidth assignments are possible at an optimal (feasible) solution. Constraint (26) states that buffer space is proportional to the equivalent capacity. Note that buffer space is assigned for all services originating at each switch, such that packets for each service may experience a delay up to the tolerated delay.

Note that we solve (24) and (26) for BW_{ijk} and B_{ijk}:

$$\begin{aligned} BW_{ijk}\left(NCMAX_{ijk}\right) &= BW : \frac{BW}{NCMAX_{ijk}} \\ &= c\left(NCMAX_{ijk}, \bar{d}_{ijk}BW, \overline{PLOSS}_{ijk}\right) \end{aligned} \tag{27}$$

Using optimality property (23) from the Theorem, the call-blocking constraint will be binding. We can solve for the the optimal arrival rate, λ_i, for a given $NCMAX_i$, using (1):

$$(28)\quad \rho^*_{ijk}\,(NCMAX_{ijk}) = \rho : \overline{BP}_{ijk} = \frac{\rho^{NCMAX_{ijk}}}{NCMAX_{ijk}!} \Bigg/ \sum_{n=0}^{NCMAX_{ijk}} \frac{\rho^n}{n!}$$

$$(29)\quad \lambda_{ijk}\,(NCMAX_{ijk}) = \frac{\rho^*_{ijk}}{T_{ijk}}$$

The arrival rate, λ_{ijk}, is determined by the limit on call-blocking. We must first calculate the maximum traffic intesity for $NCMAX_{ijk}$, given in (28), and in turn calculate the arrival rate, using (29). Because the resource allocation tables are independent of demand we have not yet related the arrival rate to a price. The bid tables, presented in the next section, will relate the price required, p_{ijk}, to produce the desired arrival rate, λ_{ijk}, as well as the marginal valuation at the volume of service provided, $NCMAX_{ijk}$.

3.3. Resource allocation between service classes. The optimality properties reduce the problem to choosing the optimal values for $NCMAX_{ijk}$. Let us we consider a linear relaxation of problem P. For solutions satisfying the optimality properties given by (23)–(26), the Karush-Kuhn-Tucker necessary conditions are trivially satisfied, with the following exception, which results from differentiating the set of constraints (19):

$$(30)\quad \frac{\partial REV_{ijk}}{\partial NCMAX_{ijk}} = \sum_{x=1}^{NS} v_x \frac{\partial BW_{ijk}}{\partial NCMAX_{ijk}} IVP_{ijkx}$$
$$1 \le i \le N, 1 \le j \le NS, 1 \le k \le NS$$

where v_x is the marginal value of capacity at switch x. Recall that IVP_{ijkx} in (34) was defined as a parameter in the problem, (P). For each service class (i,j,k), we divide both sides of (30) by $\partial BW_{ijk}/\partial NCMAX_{ijk}$, and simplify the necessary optimality conditions:

$$(31)\quad u_{ijk} = \sum_{x=1}^{NS} v_x IVP_{ijkx} \quad 1 \le i \le N, 1 \le j \le NS, 1 \le k \le NS$$

The economic interpretation of (31) is quite appealing. Any service class must yield a marginal return per unit of bandwidth equal to the sum of the marginal values for all switches (v_x for switch x) along the route the given class (i,j,k).

Incorporating the integrality requirement in $NCMAX_{ijk}$, we define marginal values per unit of bandwidth for increasing or reducing the number of connections in the solution, u^+_{ijk} and u^-_{ijk} respectively:

$$(32)\quad u^+_{ijk}\,(NCMAX_{ijk}) = \frac{REV_{ijk}\,(NCMAX_{ijk}+1) - REV_{ijk}\,(NCMAX_{ijk})}{BW_{ijk}\,(NCMAX_{ijk}+1) - BW_{ijk}\,(NCMAX_{ijk})}$$

$$u^{-}_{ijk}(NCMAX_{ijk}) = \frac{REV_{ijk}(NCMAX_{ijk}) - REV_{ijk}(NCMAX_{ijk}-1)}{BW_{ijk}(NCMAX_{ijk}) - BW_{ijk}(NCMAX_{ijk}-1)} \tag{33}$$

The marginal valuations, u^{-}_{ijk} from (32) or u^{+}_{ijk} from (33), are simply the changes in expected revenue divided by the change in bandwidth allocation, for an increase or decrease of one in $NCMAX_{ijk}$. Note that by definition, $u^{+}_{ijk}(NCMAX_{ijk})$ equals $u^{-}_{ijk}(NCMAX_{ijk}+1)$.

Discrete approximations of the continuous necessary optimalty conditions (31) are:

$$u^{+}_{ijk}(NCMAX_{ijk}) \le \sum_{x=1}^{NS} v_x IVP_{ijkx} \le u^{-}_{ijk}(NCMAX_{ijk}) \tag{34}$$
$$1 \le i \le N, 1 \le j \le NS, 1 \le k \le NS$$

In stating (34), we assume that marginal revenue is decreasing in $NCMAX_{ijk}$, i.e. $u^{-}_{ijk} > u^{+}_{ijk}$. By (34) it is not profitable to change the bandwidth allocated to any service class. The marginal value from an increased allocation is less than the sum of marginal values of bandwidth at switches along the path.

These properties dictate that optimal allocations of BW_{ijk} can be calculated from (27) based on the values of $NCMAX_{ijk}$. There is a unique arrival rates of connection requests, λ_{ijk}, associated with $NCMAX_{ijk}$, according to the call-blocking property, (28). In turn prices, p_{ijk}, are related to values of $NCMAX_{ijk}$ through the arrival rate, λ_{ijk}, given by the demand functions, (1). This gives us a reduced space of problem data, where all other problem variables are calculated as functions of $NCMAX_{ijk}$, which are integer-vlaued variables. Furthermore, we have defined marginal values of service classes, u^{-}_{ijk} and u^{+}_{ijk}, for any value of $NCMAX_{ijk}$, in (32) and (33).

We can summarize these calculations in the "Bid Table" given below:

TABLE 1
The Bid Table.

$NCMAX_i$	BW_i see (27)	B_i see (26)	λ_i see (28), (29)	p_i see (1)	u^{+}_{i} see (32)	u^{-}_{i} see (33)
1	$BW_{ijk}(1)$	$B_{ijk}(1)$	$\lambda_{ijk}(1)$	$p_i(\lambda_i(1))$	$u^{+}_{i}(1)$	$u^{-}_{i}(1)$
2	$BW_{ijk}(2)$	$B_{ijk}(2)$	$\lambda_{ijk}(2)$	$p_i(\lambda_i(2))$	$u^{+}_{i}(2)$	$u^{-}_{i}(2)$
$\vdots$	$\vdots$	$\vdots$	$\vdots$	$\vdots$	$\vdots$	$\vdots$

3.4. Estimating marginal value of bandwidth. Given the nonlinearities of the constraints, there is no elegant analytical method by which one can estimate the marginal value of capacity (bandwidth) at the switches; we use a simple bisection search procedure for this. Let

v_j^* = optimal marginal valuation for bandwidth at switch j
$\underline{v}_j$ = a lower bound on the value of v_j^* at switch j
$\bar{v}_j$ = an upperbound on the value of v_j^* at switch j

such that

$$\{\underline{v}_1, \underline{v}_2, \ldots, \underline{v}_{NS}\} < \{v_1^*, v_2^*, \ldots v_{NS}^*\} < \{\bar{v}_1, \bar{v}_2, \ldots, \bar{v}_{NS}\} \tag{35}$$

Also note that, the set of marginal values that undervalues the bandwidth at all switches results in an over-assignment of bandwidth at all switches, and a set of marginal values which overvalues the bandwidth at all switches under-assigns bandwidth at all switches. This is captured by the following relationships:

$$\sum_{i=1}^{N}\sum_{j=1}^{NS}\sum_{k=1}^{NS} BW_{ijk}\left(\underline{v}_1, \underline{v}_2, \ldots, \underline{v}_{NS}\right) IVP_{ijkx} > \overline{BW_x} \quad \forall x \tag{36}$$

$$\sum_{i=1}^{N}\sum_{j=1}^{NS}\sum_{k=1}^{NS} BW_{ijk}\left(\bar{v}_1, \bar{v}_2, \ldots, \bar{v}_{NS}\right) IVP_{ijkx} < \overline{BW_x} \quad \forall x \tag{37}$$

Our bisection search method begins with arbitrary bounds on the optimal marginal valuation of bandwidth at each switch, which satisfy (36) and (37), and by definition (35). We adjust both the upper bounds and lower bounds on marginal values simultaneously. At the upper and lower bounds of the marginal values, the amount of bandwidth that will be required for routing connections at each switch is calculated. We then perform line searches to find the revenue-maximizing feasible solution and the revenue-minimizing infeasible solution, along the line segment between the two sets of bounds on marginal values. For the bandwidth assignments given by these line searchs we adjust the marginal values of the over-valued and under-valued bandwidth. Each iteration decreses the distance between the two sets of bounds. When the upper and lower bounds are very near to each other and we have a feasible solution that fully assigns capcity at all switches, we terminate the search and give a near optimal solution, where capacity is fully assigned and revenue is optimized. The infeasible solution given by the lower bounds on marginal value of bandwidth at each switch will also provide a bound on the distance from a local optimal solution.

4. Network auction algorithm. Up to now, we have modeled the problem as though a centralized network provider was solving the entire problem in order to set prices in a situation where there are unknown connection arrival rates. At a minimum, the procedures outlined above can be used by the network provider to obtain the Bid Tables and estimates of the marginal value of capacity (bandwidth) at each switch. The problem can be easily extended to the case where there are different users at each switch bidding for the use of bandwidth for their particular class of service. Indeed, the framework that we have established allows a network operator to set up bandwidth auctions.

The network auction would proceed along the following steps:

1. *Bidding.* If each user or user agent (hereafter "agent") has been allocated circuits with bandwidth capacity for their services, then the auction is completed. Otherwise, each unassigned agent m at each switch j would announce willingness-to-pay WTP^m_{ijk} for type of service i to be switched to destination k, and would also specify the number of connections it required, NC^m_{ijk}.
2. *Preliminaries.* Calculate ΣNC^m_{ijk} and set it equal to $NCMAX_{ijk}$. If $NCMAX_{ijk}$ is infeasible (i.e. cannot be supported by bandwidth budget), go to Step 5. Otherwise, from the bid tables, determine u^+_{ijk} and u^-_{ijk}.
3. *Revising Non-Economic Bids.* If all bids of $WTP^m_{ijk} < \mathrm{u}^+_{\mathrm{ijk}}$, then it is uneconomical to provide the service. The network provider will have to announce a minimum willingness-to-accept (i.e. u^+_{ijk}). Go To Step 1.
4. *Determining Winners and Revising Economic Bids.* If some bids are > Min u^+_{ijk}, the network provider would order WTP^m_{ijk} and would accept the bid from $\mathrm{argmax}_\mathrm{m}$ $\{WTP^m_{ijk}\}$, move this agent from "blocked" to "non-blocked", and provide all the connections asked for. Announce the winner and their bid, and ask the remaining agents to revise their bids. Go to Step 1.
5. *Moderating Infeasibility.* If total demand for connections is infeasible, there could be WTP bids that are either too small or very large. If all willingness-to-pay values are large, i.e. $WTP^m_{ijk} \geq$ Max u^-_{ijk}, $\forall$m, go to Step 6. Otherwise, go to Step 3.
6. *Allocating Overbids.* Order WTP^m_{ijk}, and accept the bids from agents with the largest WTP until maximum connections are allocated. (In this case, some agents would be blocked, but revenue would be maximized.) Go to Step 1.

At the end of the auction, all bids will be between u^+_{ijk} and u^-_{ijk} which is the requirement for the marginal value of capacity at switch j. Thus, provided that the price (i.e., willingness-to-pay bid) is greater than or equal marginal value, it is economically optimal to provide the resource (i.e., bandwidth for a particular class of service). Further, resources provided to all connections can be met within the bandwidth budget. In the case of very low initial bids, there is a good chance that the agents can get away eventually with the minimum bid required for service. In the case where the initial bids are very high, some users will get "blocked".

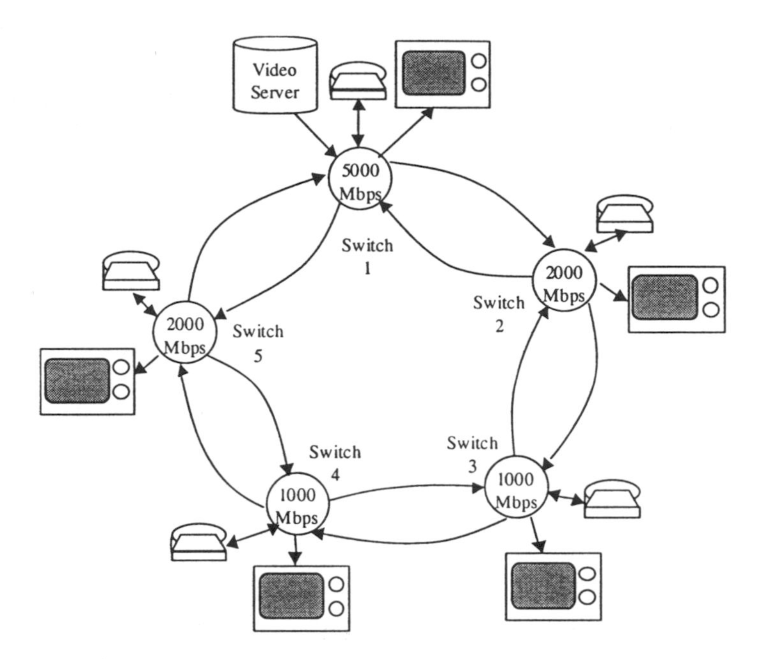

FIG. 2. *Two service network with a single video server.*

5. Example problem. We consider a bi-directional ring network offering voice and video connection to users via a set of five switches. The example network and available services are illustrated in Figure 2 below. The voice connections can be either local, i.e., routed through a single switch, or long-distance to anywhere in the network, i.e. routed from any origin to any destination switch. The video connection originates from a single switch (labeled "1" in the figure), where a video server is available and may be routed to any switch in the network, including the originating node. The connections are all routed using the minimum hop routing. The service class definitions are given in Table 2 that also provides traffic data.

TABLE 2
Service class definitions.

	Data Traffic Parameters			**Other Parameters**			
Service Class	Mean. Rate (Mbps) r_i	Peak. Rate (Mbps) R_i	Average Burst (s) b_i	Packet Loss Probability $\overline{PLOSS_i}$	Blocking Probability $\overline{BP_i}$	Delay (s) $\bar{d_i}$	Average Holding Time (s) T_i
Voice	0.032	0.064	1.0	10.0×10^{-5}	0.01	0.0	3.0
Video	1.000	10.000	10.0	10.0×10^{-9}	0.01	5.0	30.0

We assume constant elasticity of demand for both services, with the elasticity values taken from [9] and presented in Table 3.

TABLE 3
Demand for services.

Service Class	Demand
Voice ($\forall j, k$)	$\lambda_{voice,jk} = 1.0p^{-1.05}$
Video ($j = 1$)	$\lambda_{video,jk} = 2.0p^{-2.50}$
Video ($j \neq 1$)	0

Suppose we solve the problem of setting prices at a centralized problem using the methodology given above. The steps involve estimating marginal values of capacity at each switch, using the bid tables to estimate the maximum number of connections supported by this marginal value, and then obtaining prices from the inverse function of the arrival rates. (For details, see [6].) The optimal prices are given in Table 4. Note that at these prices, the bandwidth assignments at every switch are in excess of 99% and revenue is maximized.

TABLE 4
Optimal price per unit time for a connection of service (i, j, k).

Service Type	Origin	Destination (k)				
(i)	(j)	1	2	3	4	5
Voice	1	0.76	0.88	1.06	1.06	0.88
	2	0.88	0.10	0.24	0.40	1.00
	3	1.06	0.24	0.14	0.29	0.40
	4	1.06	0.40	0.29	0.14	0.24
	5	0.88	1.00	0.40	0.24	0.10
Video	1	1.65	1.91	2.32	2.32	1.91

TABLE 5
Optimal price per unit time for a connection of service (i, j, k).

Service Type	Origin	Destination (k)				
(i)	(j)	1	2	3	4	5
Voice	1	0.10	0.50	0.60	0.60	0.50
	2	0.50	0.10	0.20	0.40	1.00
	3	0.75	0.20	0.10	0.25	0.40
	4	0.75	0.40	0.25	0.10	0.20
	5	0.50	0.60	0.40	0.20	0.10
Video	1	2.50	3.00	3.00	3.00	2.50

6. Concluding remarks. The network was also used to assess how an auction mechanism could be implemented using experiments in a graduate course in systems engineering at the University of Maryland. The students were given the characteristics of the traffic data and the network

(Table 2), and were asked to bid for service. The experiment was conducted for 10 sessions with different sets of students. The average results are given in Table 5. The agents always under-bid for voice services and made reasonable bids for video services. Eventually, the voice bids (and hence agreed upon prices) were the minimum marginal cost (i.e. minimum willingness-to-accept) announced by the network operator, and lower than the optimal prices shown in Table 4. On the other hand, bids for video services were much higher than that obtained through centralized evaluation of optimal prices.

In this paper, we have modeled the problem of determining prices for multiple telecommunications services as a nonlinear integer program. The objective is to maximize the expected revenue for the network operator (or service provider). The constraints ensure that quality of service guarantees are met by limiting the delay and dropped packets to be below pre-set limits for the different service classes. The network operator would also like to limit the number of calls (or service requests) that are blocked. The problem was solved from the perspective of a centralized operator, and using some structural results from this, as an auction of bandwidth. In both cases, the "bid table" was used to determine the best allocation of maximum connections of a particular class of service and the marginal values of either adding or dropping connections. Numerical results showed that one could optimize prices and resource allocations for a centralized network operator. However, using an auction mechanism, one could get very different prices and may also be in a position to increase revenues, provided the call blocking constraint was relaxed.

REFERENCES

[1] Cocchi, R., S. Shenker, D. Estrin and L. Zhang, "Pricing in Computer Networks: Motivation, Formulation, and Example," IEEE/ACM Transactions on Networking, vol. 1, no. 6, 614–627 (1993).

[2] Guérin, R., H. Ahmadi and M. Naghshineh, "Equivalent Capacity and its Application to Bandwidth Allocation in High-Speed Networks," IEEE Journal on Selected Areas in Communications, vol. 9, no. 7, 968–981 (1991).

[3] Gibbens, R.J. and F.P. Kelly, "Resource Pricing and the Evolution of Congestion Control," draft paper (1999), (available at http://www.statslab.cam.ac.uk/~frank/evol.html).

[4] Jiang, H. and S. Jordan, "The Role of Price in the Connection Establishment Process," European Transactions on Telecommunications — Economics of Telecommunications, November 9 (1994).

[5] Keon, N.J. and G. Anandalingam, "A New Pricing Model for Competitive Telecommunications Services Using Congestion Discounts", Working Paper, University of Pennsylvania, July 2000 (Submitted for Publication).

[6] Keon, N.J. *Pricing of Multiple Services in Telecommunications Networks*, Ph.D dissertation, Department of Systems Engineering, University of Pennsylvania, May 2000.

[7] Kelly, F.P. A.K. Maullo and D.K.H. Tan, "Rate Control in Communication Networks: Shadow Prices, Proportional Fairness and Stability," Journal of the Operational Research Society, vol. 49, 237–252 (1998), (available at http://www.statslab.cam.ac.uk/~frank/rate.html).

[8] Korilis, Y.A., T.A. Varvarigou and S.R. Ahuja, "Pricing Noncooperative Networks," submitted to the IEEE/ACM Transactions on Networking, May 1997.
[9] Lanning, S., D.Mitra, Q. Wang and M. Wright, "Optimal Planning for Optical Transport Networks," presented at the Fifth INFORMS Telecommunications Conference, Boca Raton, FL, March 5–8, 2000.
[10] Low, S.H. and P. Varaiya, "A New Approach to Service Provisioning in ATM Networks," IEEE/ACM Transactions on Networking, vol. 1, no. 5, 547–553 (1993).
[11] Low, S. H., "Equilibrium Bandwidth and Buffer Allocations for Elastic Traffics," Submitted for publication, May 1997.
[12] Mackie-Mason, J.K. and H. Varian, "Pricing the Internet," Public Access to the Internet, eds. B. Kahin and J. Keller, Cambridge and London: MIT Press, 269–314, 1995.
[13] Mackie-Mason, J.K. and H. Varian, "Some Economics of the Internet," in: Networks, Infrastructure and the New Task for Regulation, eds. W. Sichel and D.L. Alexander, Ann Arbor: University of Michigan Press, 107–36 (1996).
[14] Masuda, Y. and S. Whang, "Dynamic Pricing for Network Service: Equilibrium and Stability," Management Science, vol. 45, no. 6, 857–869 (1999).
[15] Thomas, P., Teneketzis, D. and Mackie-Mason, J.K., "A Market-Based Approach to Optimal Resource Allocation in Integrated-Services Connection-Oriented Networks," submitted for publication, August 1999.
[16] De Veciana, G. and R. Baldick, "Resource Allocation in Multi-Service Networks via Pricing: Statistical Multiplexing," Computer Networks and ISDN Systems, vol. 30, 951–962 (1998).
[17] Wang, Q., J.M. Peha, and M.A. Sirbu, "The Design of an Optimal Pricing Scheme for ATM Integrated Services Networks," Special Issue: Internet Economics, Journal of Electronic Publishing, University of Michigan Press, vol. 2, no. 1 (1996).

PRICEBOT DYNAMICS

SUZHOU HUANG*, KEVIN ANDERSON†, AND YONG YANG*‡

Abstract. E-commerce through the Internet has created a special economic environment in which transaction costs and search costs are substantially lower. Taking advantage of this new situation software intermediaries naturally appear and even become pervasive. Shopbots that search and collate information for buyers on the Internet are such examples. It is envisioned that similar computer agents that act on behalf of sellers will soon emerge. Pricebots that dynamically set prices for online sellers are expected to be in the next generation of software intermediaries. In this paper we study a class of sequential duopoly pricing games that models some aspects of pricebot dynamics. Under Markov settings these games can be solved via backward induction. The solution is found to display very complex patterns as a function of the discount factor, due to bifurcation phenomena in the discrete map induced by backward induction. However, it is possible to define an effective but simpler dynamics that retains the optimality of the original game. We further show that this effective dynamics can be sustained by steady self-confirming equilibria. Our results (1) set limits on what learning algorithms based on Markov assumptions can achieve and (2) imply that learning in this kind of game should not be focused on the exact reaction functions, but rather on achieving optimal net present values with the realized time series of prices.

1. Introduction. Selling goods on the Internet typically requires sellers to post a variety of information such as prices online. Due to negligible search costs comparison shopping becomes much easier on the net. Taking advantage of the readily available information, software agents that search and collate information for buyers on the Internet are now widespread. There are many such examples of shopbots: BargainFinder, DealPilot, mySimon, Shopper, Jango and Frictionless to name a few. In response to this kind of e-commerce environment, sellers are likely to deploy computer agents of their own. Pricebots that dynamically set prices for sellers by explicitly taking their competitors' prices into account are expected to emerge soon. Of course, the purpose of designing a good pricebot is to make profit in the long run by sustaining a healthy balance of market share and price. To achieve that goal some kind of optimal pricing strategies must be encoded in pricebots. One aspect that poses a particular challenge for pricebot design is that not all relevant information are revealed on the Internet. For example, competitors' cost structures and decision-making processes may be hidden. In order to perform well, pricebots need to have the capability to automatically adapt their strategies to (or learn from) the observed actions of their competitors.

Among the first attempts that address how pricebots behave in the emerging e-commerce environment was the work of an IBM group, Green-

*Ford Research Laboratory, Ford Motor Company, Dearborn, MI 48121-2053. shuang10@ford.com.

†Institute for Mathematics and its Applications, University of Minnesota, Minneapolis, MN 55455. kanderso@ima.umn.edu.

‡yyang1@ford.com.

wald, Kephart and Tesauro (1999) and Tesauro and Kephart (1999, 2000), which examined a stylized model that captures some of the elements in the envisioned context. While achieving limited successes, their learning algorithm also experienced some difficulties in converging to the desired solution when sellers are sufficiently patient. In order to have a clear assessment of what is really going on it is desirable to have a quantitative characterization of the pertinent equilibrium structure in the kind of games they considered. This quantitative assessment will not only help in understanding why a particular learning algorithm fails, but also serve as a bound that a learning algorithm can ever achieve. Perhaps more importantly, the equilibrium structure, once elucidated properly, can provide a conceptual guidance in designing future learning algorithms. Furthermore, if the structure of the equilibrium is too complicated to learn näively, it may become imperative to develop more elaborate learning criteria. Our work summarized in the following is meant to be such a pre-learning exploration.

The economics of the IBM model turns out to belong to the class of games considered by Maskin and Tirole (1988). This class is called sequential duopoly pricing games with perfect information. When the strategy space is limited to Markov, Maskin and Tirole (1988) show that Markov subgame perfect equilibria (MPE) are solutions to a set of coupled Bellman equations. However, solving these equations in general is a formidable task. Our contribution in this paper essentially rests on the observation that backward induction (BI) can be used to numerically find equilibria of these games. Usually, backward induction requires a finite time horizon and a given terminal condition, whereas our main interest is really at the long run limit. As we will demonstrate, backward induction can be viewed as a time-reversed map, which we call the BI-map. If this map possesses any invariant set of ω-limit points that is finite dimensional, such an ω-limit set constitutes a bona fide MPE of the game with an infinite time horizon. Therefore, our aim here is to obtain at least some MPEs of the game by studying the long time limit of the corresponding BI-map, and then examine how these MPEs depend on the exogenous parameters. Lastly, we explore the implication of these MPEs to learning in this class of games.

2. The model. Since we have learning in mind, we specifically consider the model proposed by the IBM group. Qualitatively similar results are expected for other games in this class. In this model, there are two sellers and infinitely many consumers. Both sellers sell identical goods that are perishable. Time is measured in discrete periods. Sellers dynamically respond to each others' prices. This is modeled by allowing them to take turns setting prices and insisting that each chosen price is committed for two time periods. For concreteness, we assume that firm a sets prices in odd periods and firm b in even periods. Consumers are represented by two kinds of agents. The simple-minded agent randomly chooses a seller from whom to buy a good. The sophisticated agent always buys from the seller with the lowest price. $w \in (0,1)$ parameterizes the fraction of consumers who are represented by the simple-minded agent. w will be fixed at 0.25 throughout the paper. The resulting payoff (normalized) per period, say for firm a, is given by

$$\Pi_a[p_a, p_b] = \begin{cases} (1 - w/2)\, p_a, & \text{if } p_a < p_b \\ (w/2)\, p_a, & \text{if } p_a > p_b \\ p_a/2, & \text{if } p_a = p_b \end{cases} \tag{1}$$

where prices can take N discrete values $\mathcal{P} = \{p_i = i/N; i = 0, 1, \cdots, N-1\}$. The ceiling on p_i represents the assumption that no consumer is willing to pay more than 1 for the good. $\Delta p = 1/N$ is the minimum amount that a firm can under-cut the price set by its competitor, in order to induce appreciable demand changes. So, $\Delta p = 1/N$ (or N) should be regarded as one of the exogenous parameters of the model, rather than a measure of discretization. Both firms are assumed to only play Markov strategies. The economics of this game is intuitively simple: both firms wish to maintain a good balance between market share and high prices so that profits can be sustained in the long run.

According to the theorem of Zermelo and Kuhn (see Fudenberg and Tirole (1991)), for any finite extensive game and a given terminal condition, if there is no indifference in the induction process the equilibrium thus attained is unique subgame perfect Nash. Because of the non-degenerate nature of Eq.(1), BI provides us a numerical means to totally characterize the equilibrium using pure strategies for each given terminal condition, perhaps barring a set of accidental cases of measure zero. Consideration of mixed strategies will not be necessary.

Let V_a and W_b denote the net present values (NPVs) of firm a and b at odd periods when firm a is picking new prices, and $\rho \in (0,1)$ the common discount factor. Note that V_a and W_b are N-dimensional vectors labeled by the state variable $p_i \in \mathcal{P}$. Then the BI process is defined by a pair of Bellman equations (see Maskin and Tirole (1988) for details), for every $p_i \in \mathcal{P}$,

$$\begin{aligned} V_a^t[p_i] &= \Pi_a[R_a^t(p_i), p_i] + \rho\, W_a^{t+1}[R_a^t(p_i)]\,, \\ W_b^t[p_i] &= \Pi_b[p_i, R_a^t(p_i)] + \rho\, V_b^{t+1}[R_a^t(p_i)]\,, \end{aligned} \tag{2}$$

where $R_a^t(p_i)$ is firm a's reaction function (RF). This function provides the price firm a will set in response to firm b having set price p_i. It is given by a maximization process

$$R_a^t(p_i) = \arg\max_{p_j \in \mathcal{P}} \Big\{ \Pi_a[p_j, p_i] + \rho\, W_a^{t+1}[p_j] \Big\}\,. \tag{3}$$

A similar process is defined for V_b and W_a at even periods when firm b chooses new prices. Solutions to these equations are automatically Markov subgame perfect Nash equilibria (MPE).

Eqs.(2,3) can be viewed as two N-dimensional time-reversed maps (differing only in offsets) in the space spanned by the vectors W_a and V_b. This can be made explicit by iterating the original map once, and hence directly relating W_a^{t-1} to W_a^{t+1} and V_b^{t-1} to V_b^{t+1},

$$
\begin{aligned}
W_a^{t-1} &= \Pi_a^{t-1} + \rho\,\Pi_a^t + \rho^2\, M\, W_a^{t+1}\,, \\
V_b^{t-1} &= \Pi_b^{t-1} + \rho\,\Pi_b^t + \rho^2\, M\, V_b^{t+1}\,,
\end{aligned}
\tag{4}
$$

where the two immediate payoff terms, Π^{t-1} and Π^t, do not have explicit dependence on W_a and V_b. The matrix M represents the mapping sequence $R_a^t(R_b^{t-1})$, with R_b^{t-1} given by firm b's maximization process similar to Eq.(3), i.e., $M_{ij} = 1$ if $R_a^t(R_b^{t-1}(p_i)) = p_j$ and $M_{ij} = 0$ otherwise. However, due to this step of iteration R_b^{t-1} becomes R_a^t dependent. This kind of nested RFs of different players distinguishes game environments from the value iteration in control environments, where there is only one active player. The terminal condition W_a^T and V_b^T of the BI process now play the role of the initial condition for the BI-map. The control parameter of the map is the discount factor ρ. It is conceptually important to realize that RFs, which only take discrete values, are not the mapping variables, but rather some intermediate objects that facilitate the mapping.

Although much of our explicit results are numerical, many things can be said about the BI-map analytically. First, Eq.(3) can be used to partition the N-dimensional space spanned by W_a into disjoint regions. Each region, which we will call a stability region, is the set of W_a's for which the solution to Eq.(3) is a fixed reaction function. These regions are bordered by simple hyper-planes with slopes either being 0 or 1 and intersections specified by the differences between pairs of payoff Π_a's. So, the BI-map can almost be regarded as a multiple dimensional generalization of the tent map on an interval (with the mapping function $f(x) \propto |x - 1|$ for $x \in (0, 2)$). Rand (1978) has demonstrated the existence of chaos in a version of Cournot duopoly game with alternating-move, best-response dynamics for myopic players modeled by two nested tent maps (see also Dana and Montrucchio (1986)).

Second, within any such a region where the RF is fixed, the BI-map is linear and can be shown to be a local contraction. This property can be inferred from the once iterated BI-map Eq.(4), by recognizing that M is a degenerate permutation matrix with eigenvalues either 0 or modulus 1. This argument requires that both R_a^t and R_b^{t-1} be such that W_a^{t+1} and V_b^{t+1} remain in one of their stability regions. Cycles, due to global bifurcation, will occur if it is not possible to maintain this requirement.

Since our goal is to understand the long-run limit of the game, we are interested in the invariant set of ω-limit points (see Hirsch and Smale (1974)) of the BI-map. To this end, we will take a very large value for T and focus our attention on equilibrium behavior at $t \ll T$. If W_a^t and V_b^t reach a fixed point, i.e. settle to a cycle with periodicity 1 (or 2) at small

t, this cycle is identified as a symmetric (or asymmetric) time-independent MPE. This time-independent MPE can be understood as the one defined in the same game with an infinite time horizon. If W_a^t and V_b^t settle to a cycle with periodicity greater than 2 (aperiodic mapping itineraries are viewed as periodicity=∞), then time-independent MPE does not exist for the given terminal condition. In the infinite time horizon context, these cycles with periodicity greater than 2 consist of a new class of solutions to the game in which the NPV and RF vectors are periodic functions of time. Maskin and Tirole (1988) did not explicitly address these "cyclic" MPEs in their paper. Of course, different terminal conditions may converge to the same cycle, or lead to totally different itineraries. This non-trivial dependence on the terminal condition is consistent with the fact that there are multiple equilibria in this class of games, as shown by Maskin and Tirole (1988) and verified in our numerical calculations. Similar dependence on the initial condition has been rigorously shown to arise in the 2-dimensional Hénon map (see Newhouse (1979)). In this case there exist basins of attraction to many different stable cycles. In the rest of the paper we will only present result with the zero terminal condition: $W_a^T = 0$ and $V_b^T = 0$, representing the economic scenario that the selling stops at $t \geq T$.

3. Result. For relatively small value of ρ, the BI process usually converges to a symmetric solution from the zero terminal condition. How small the value of ρ has to be in order to have this convergence is N-dependent (see Figure 3 for detail). An example of RF, corresponding to a symmetric MPE at $N = 100$ and $\rho = 0.8$, is given in Figure 1. The realized prices using this RF is a pricing war that never settles. Both firms keep undercutting each other until a price is reached where it is beneficial to make short-run sacrifice and return to the top of the cycle (p_{N-1}). Behaviors like this are termed as Edgeworth cycles by Maskin and Tirole (1988). We also observe, though not with the zero terminal condition, equilibrium behaviors that are called Kinked-demand curves, in which the pricing war is temporary and settles to a fixed price.

As ρ gradually increases, asymmetric MPEs (cycles of periodicity 2) appear. Further increasing ρ will result in much more complicated cyclic patterns. It is important to note that these cycles are parts of MPEs of the game, not those appearing in any particular learning dynamics, such as the Shapley cycles in various forms of fictitious play. To have a quantitative feeling of these patterns, let us focus on a relatively small system: $N = 11$. In order to detect the relevant periodicity, we introduce a fictitious observable $\widetilde{\mathrm{NPV}}_t$, representing the total NPV averaged over all states (p_i) and over both players at period t. In Figure 2, we display $\widetilde{\mathrm{NPV}}_t$ as a function of ρ for $t < T - 1000$. We skip the first 1000 iterations to minimize the effect of transience inherent from the terminal condition. This is analogous to the bifurcation diagram typically given for one-dimensional maps, such as the logistic map (see Collet and Eckmann (1980)). The multiplicity at a

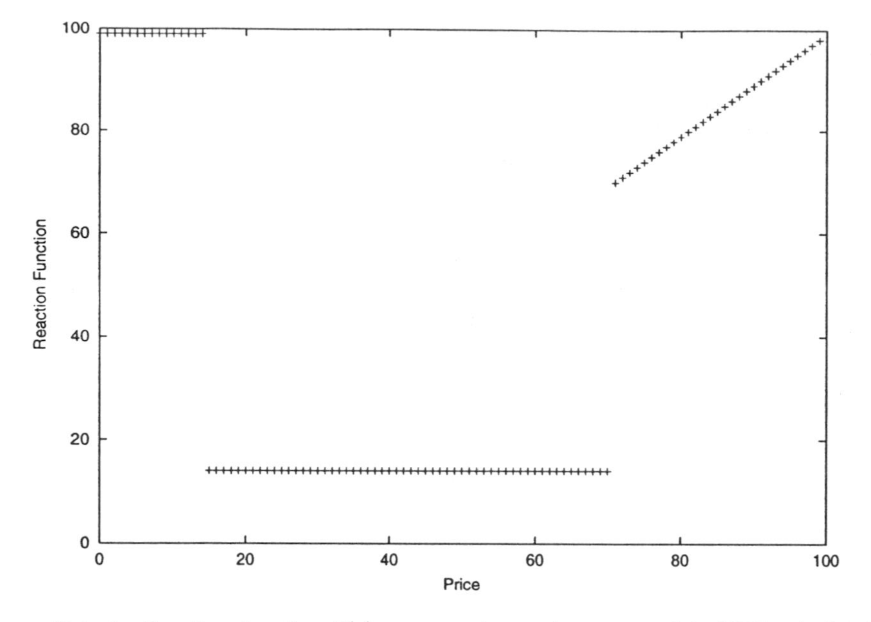

FIG. 1. *Reaction function R(p) versus price, p, in a symmetric MPE calculated from the BI process with $N = 100$ and $\rho = 0.8$.*

given ρ indicates the maximum number of distinct RFs in the corresponding cycle. For example, there are three distinct RFs at $\rho = 0.94$, though the periodicity is six at the same value of ρ. To exemplify the complexity of the bifurcation patterns the inset in Figure 2 shows the detailed structure for $\rho \sim 0.30823$. Our non-exhaustive exploration seems to indicate that there are no aperiodic itineraries in the BI-map, though we do observe quite long periodicities (for example, 18 in the case of $N = 11$ and $\rho = 0.3083$). We do not know whether the lack of aperiodic itineraries in our numerical study is accidental, the set of initial condition that leads to aperiodic itineraries having measure zero or the true absence. (For a relatively recent review of bifurcations and chaos in other economics contexts see Baumol and Benhabib (1989), which relies on analogies to the intuitive arguments of May (1976) in the logistic map.)

As pointed out earlier, bifurcations occur in the BI-map when the NPVs are mapped from stability region of one RF to that of another. Longer cycles correspond to the situation where NPVs are mapped through a periodic sequence of stability regions. A cycle of periodicity k can also be viewed as a fixed point of the $(k-1)$-time iterated BI-map, described by a $(k \times N)$-dimensional Bellman equation. This equation can be obtained by embedding Eqs.(2,3) in block forms into $k \times N$ dimensions sequentially.

Our result immediately lends an explanation to the non-convergence of the Q-learning algorithm for large ρ observed in Kephart and Tesauro (1999, 2000). Once the value of ρ can sustain a cycle with periodicity

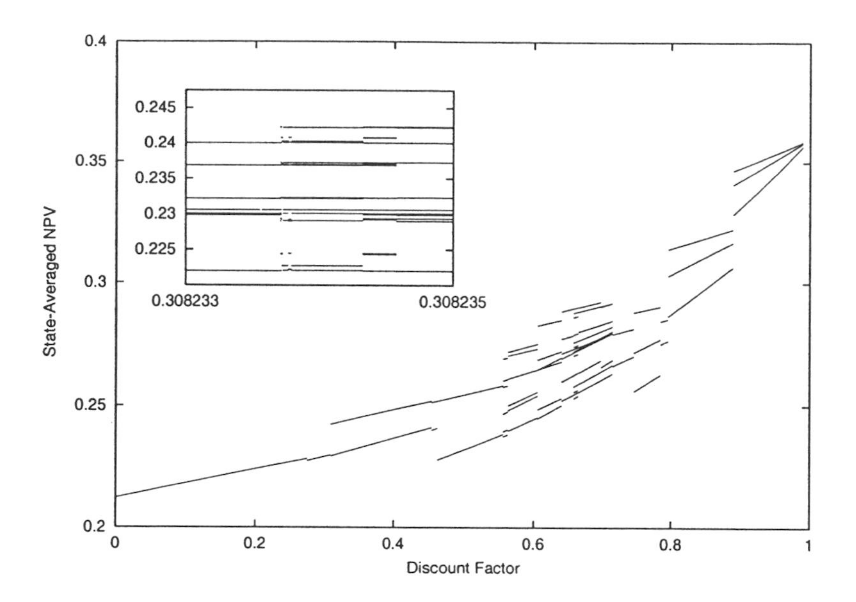

FIG. 2. *The state-averaged net present value, $\widetilde{NPV}_t$ (for all $t < T - 1000$), versus the discount factor ρ. The inset shows the fine structure for $\rho \sim 0.30823$.*

greater than 1, the Q-function is no longer time independent, even though the system is autonomous. It is likely that the critical value ρ_c beyond which bifurcation phenomena appear depends on the details of the learning algorithm, such as the initial condition and updating procedure. As an example, in Figure 3, we plot ρ_c as a function of N, where we define ρ_c to be the discount factor above which the symmetric MPE can not be reached from the zero terminal condition.

Due to the complexity of the cycle patterns, it appears that it is not possible for any learning algorithm to learn the entire sequence of the RFs in the cycle. Furthermore, the exact RFs may not be robust when the exogenous parameters are slightly varied. These would be rather discouraging statements, if we could not find something else that captures the essence of the dynamics, on the one hand, and is simpler, robust and still maintains some kind of optimality on the other. Fortunately, the situation is not so dire, at least in this model. For a given sequence of reaction functions, $R_a^t(p_i)$ and $R_b^{t-1}(p_i)$, we can define the realized time series of prices as $p^t = R_a^t(p^{t-1})$ and $p^{t-1} = R_b^t(p^{t-2})$ for odd and even time periods respectively. If we focus our attention on these realized prices, the dynamics are much simpler. For one, even in the case of a set of parameter values for which the BI-map has high periodicity, the number of prices in the realized price cycle can be quite small. To detect this effective dynamics, all what we need to do is to histogram the time series of the realized

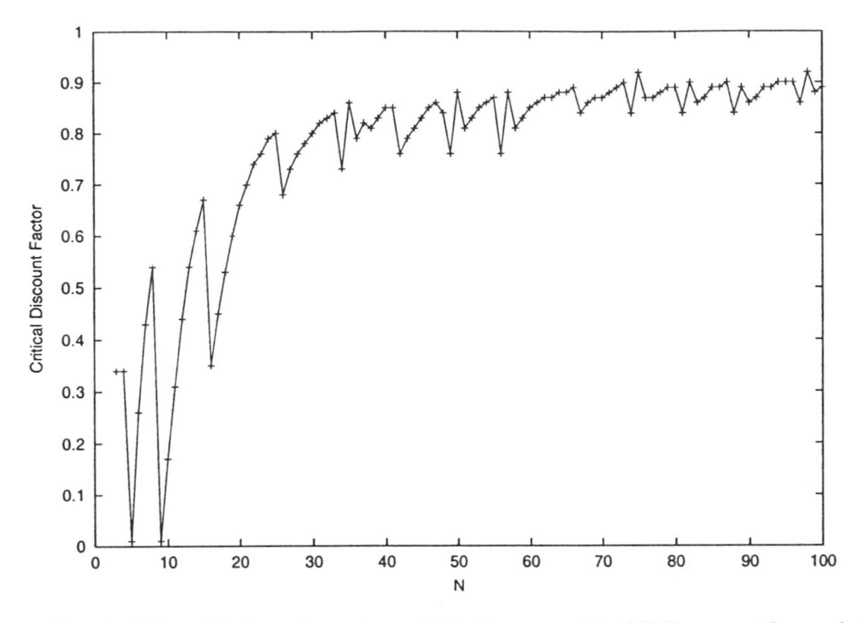

FIG. 3. *The critical ρ value, above which the symmetric MPE can not be reached from the zero terminal condition, versus the number of price points N.*

prices. Only those points that repeat in the cycle will have non-vanishing entries at the end. Let us denote the collection of these price points by $\bar{\mathcal{P}} \subseteq \mathcal{P}$. It is then not hard to see that the effective RFs on the set $\bar{\mathcal{P}}$ are time independent. When the number of points in $\bar{\mathcal{P}}$ is odd (even), an effective symmetric (asymmetric) MPE emerges. Furthermore, the NPVs evaluated on $\bar{\mathcal{P}}$ coincide with those defined in the BI process. Therefore, the effective dynamics retains the full optimality of the original game. We can further write down the effective Bellman equation, which amounts to finding an invariant subset of the $(k \times N)$-dimensional Bellman equation for a cycle of periodicity k. This invariant subset is defined on $\bar{\mathcal{P}}$, hence its dimensionality is reduced. This reduction in dimensionality can be substantial, especially for large values of ρ, where the economic outcomes are found to be more collusive. For example, when $N = 11$ and $\rho = 0.94$, we have $k = 6$, $\bar{\mathcal{P}} = \{p_7, p_9, p_{10}\}$. The symmetric effective RF is given by $\bar{R}(p_{10}) = p_9$, $\bar{R}(p_9) = p_7$ and $\bar{R}(p_7) = p_{10}$. Those components of NPV vectors defined on $\bar{\mathcal{P}}$, while obeying a simple form of Bellman equation, can also be calculated directly by summing over discounted payoffs evaluated on an infinite price sequence with the ordering specified by $\bar{R}$. These components consist of an ergodic class of a Markov chain induced by $\bar{R}$ on $\mathcal{P}$. According to Maskin and Tirole (1988), a singleton ergodic class corresponds to a kinked demand curve MPE and a non-singleton ergodic class corresponds to an Edgeworth cycle MPE.

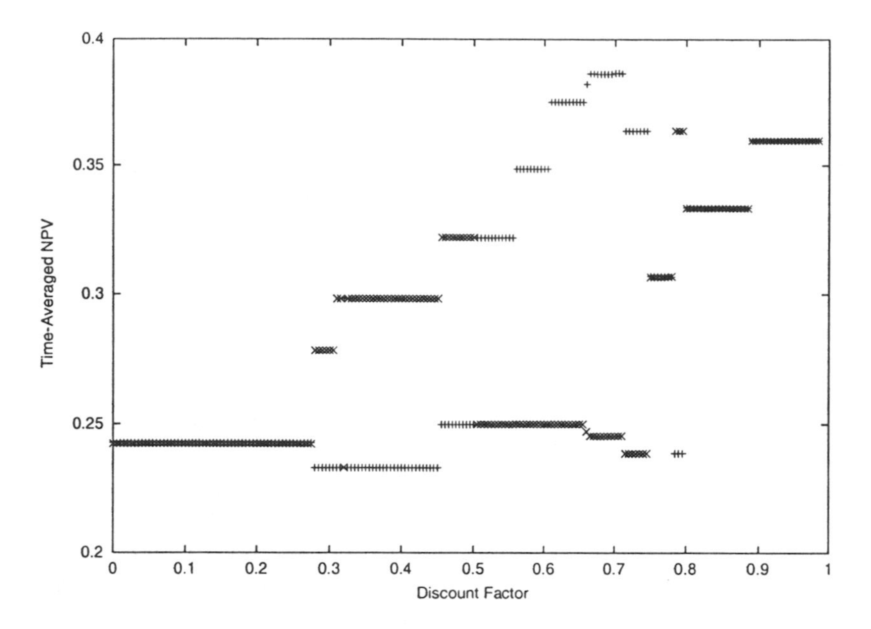

FIG. 4. *The time-averaged net present value,* $\overline{NPV}$, *versus the discount factor* ρ. *The x's are for firm a and +'s for firm b. The asymmetry for some values of* ρ *between x's and +'s is due to the fact that the number of points in* $\bar{\mathcal{P}}$ *is even for those values of* ρ, *and that the order of moves is inherently asymmetric.*

In Figure 4 we show the ρ dependence of the effective equilibrium, where the vertical axis is $\overline{\text{NPV}}$, defined to be the time-averaged NPV in the cycle for each player separately. This definition is analogous to invariant distributions in one-dimensional maps. It is evident that $\overline{\text{NPV}}$ only receives contribution from the original NPVs on the set $\bar{\mathcal{P}}$. This is partially reflected from the figure by the fact that $\overline{\text{NPV}}$ remains constant and jumps only when the effective RF shifts. Single-valuedness (double-valuedness) indicates that the effective MPE is symmetric (asymmetric). Contrasting Figures 2 and 4, we are truly amazed that the complex bifurcation structure in "cyclic" MPEs of the game merely facilitates a simple effective dynamics that is qualitatively similar to an Edgeworth cycle of a time-independent MPE. The gaps in $\overline{\text{NPV}}$ at those values of ρ where the effective dynamics is asymmetric are expected to shrink when N becomes large.

4. Implication to learning. In order to carry out the BI process, all information must be common knowledge. In a practical economic environment, and especially that of e-commerce, the assumption of common knowledge is highly questionable. Furthermore, it is desirable that the learning can be done quickly and simultaneously with business practice. This means that the learning algorithm should be designed so that it is trainable on-line and its computational demand is not worse than linear in N. Because of these reasons, the learning process can at best be a self-confirming process. (see Fudenberg and Levine (1998) for a fuller description on the concept of self-confirming). In these processes, players need to start with some prior beliefs, and then choose actions and update beliefs concurrently as the learning process unfolds. Our results indicate that, while it is almost hopeless to learn the exact RF, due to complex bifurcation patterns, the attention of learning should be focused on the effective dynamics of the game. This emphasis in achieving optimal NPVs on $\bar{\mathcal{P}}$ is close in spirit to what was called the ϵ-universal consistency by Fudenberg and Levine (1998) in the context of fictitious play. Because $\bar{\mathcal{P}}$ can be a much smaller set than $\mathcal{P}$, requirements like self-confirming and ϵ-universal consistency may be easier to meet. In fact, it is not hard to see that the effective dynamics on $\bar{\mathcal{P}}$ for the example (with $N = 11$) we considered earlier can be sustained by steady self-confirming equilibria, as long as prior beliefs are such that NPVs on $\mathcal{P}\backslash\bar{\mathcal{P}}$ are sufficiently small and RFs take values only in $\bar{\mathcal{P}}$.

Finally, another desirable feature of a good learning algorithm is that there is no implied structure for the solution to the game it is trying to learn. For example, one should not assume a priori that the reaction function is time independent. The insight gained via BI in this paper indeed motivates an algorithm that apparently satisfies all these requirements. Whether this algorithm will perform well and how it is related to other learning algorithms already in the literature will be explored systematically in the near future.

REFERENCES

[1] Greenwald, A.R., Kephart, J.O., and Tesauro, G.J., *Proceedings of the First ACM Conference on Electronic Commerce*, 1999, ACM press.

[2] Tesauro, G.J. and Kephart, J.O., *Proceedings of Workshop on Game-Theoretic and Decision-Theoretic Agents*, 1999, University College London, in press.

[3] Kephart, J.O. and Tesauro, G.J., *Proceedings of Seventeenth International Conference on Machine Learning*, 2000, Stanford University, in press.

[4] Maskin, E. and Tirole, J., *Econometrica*, **56**, 571–599 (1988).

[5] Fudenberg, D. and Tirole, J., *Game Theory*, (The MIT Press, Cambridge, Massachusetts, 1991).

[6] Rand, D., *Journal of Mathematical Economics*, **5**, 173–184 (1978).

[7] Dana, R.A. and Montrucchio, L., *Journal of Economic Theory*, **40**, 40–56 (1986).

[8] Newhouse, S.E., *Publ. Math. IHES*, **50**, 101–151 (1979).

[9] Collet, P. and Eckmann, J.P., *Iterated Maps on the Interval as Dynamic Systems*, (Birkhäuser, Boston, 1980).

[10] Baumol, W.J. and Benhabib, J., *Journal of Economic Perspectives*, **3**, 77–105 (1989).

[11] May, R., *Nature*, **261**, 459–467 (1976).

[12] Hirsch, M. and Smale, S., *Differential Equations, Dynamic Systems, and Linear Algebra*, (New York, Academic Press, 1974).

[13] Fudenberg, D. and Levine, D. K., *The Theory of Learning in Games*, (The MIT Press, Cambridge, Massachusetts, 1998).

THRESHOLD PRICING FOR SELLING NETWORK CAPACITY THROUGH FORWARD CONTRACTS

STEVE G. LANNING*, WILLIAM A. MASSEY†, AND QIONG WANG‡

Abstract. In this paper, we consider a telecommunications carrier that serves end users and sells capacities to its peers. We formulate a feedback procedure by which the carrier lets customers bid on capacity contracts and dynamically sets threshold prices to determine which bids to accept. The approach maximizes the carrier's total revenue from both service and capacity markets. We develop key mathematical techniques for calculating the optimal threshold prices and discuss their properties.

Key words. Bid acceptance, pricing, network capacity, Erlang blocking formula.

AMS(MOS) subject classifications. 91B26, 91B70, 60K26, 90B22, 91B24.

1. Introduction. Network capacity trading among carriers is becoming an industry trend in the telecommunications sector. Efforts have been made to set up exchange markets where carriers sell their capacities as commodities. In other cases, capacity exchange is carried out through forward contracts between interested parties, see Wischik and Greenberg [12] for example.

In this paper, we consider carriers that are only interested in selling capacities. Each carrier decides how to divide its capacity between using some for serving the end users and the rest for exchanging with other carriers. For many carriers, serving end-users is the "bread-and-butter" of their business. These customers arrive to the network at random, acquire a communication channel for a certain period of time, and pay a price based on the duration of the service. We call the revenue from end users *service revenue.* In the presence of capacity exchange, carriers get additional revenue by selling capacities to others, but doing so reduces their ability to serve end-users, which affects their service revenue. It is in a carrier's best interest to sell capacities only when the value derived from selling can justify the loss of service revenue.

The value derived from selling network capacities depends strongly on the way the trading is performed. Carriers can auction their capacity to others, in which case the design of auction protocols is important. Carriers can also sell capacities in an exchange market operated by a third party, where they can apply various financial instruments, such as options and futures contracts, to maximize their gains from trading. Furthermore, each

*Aerie Networks, 1400 Glenarm Place, Denver, CO 80202; Email: slanning @aerienetworks.com.

† Bell Labs, Lucent Technologies, 600 Mountain Avenue, Murray Hill, NJ 07974; Email: wmassey@lucent.com.

‡Bell Labs, Lucent Technologies, 600 Mountain Avenue, Murray Hill, NJ 07974; Email: chiwang@lucent.com.

carrier can sign a forward contract with another carrier, where the revenue from selling capacity depends on the terms of the contract. All these possibilities cannot be addressed in a single paper, so we limit our study here to the case of a carrier selling its capacity by forward contracting. These contracts specify the amount of network capacity that can be used, as well as the starting and ending times for usage. Interested parties can bid on price. The carrier dynamically sets an unannounced threshold price above which the bid will be accepted. We discuss how to set this threshold price optimally so the carrier can maximize its revenue from both serving end-users and selling capacities to other carriers.

Our problem falls into the general category of revenue management, where a vast amount of research has been performed for the airline industry (for a comprehensive review, see McGill and van Ryzin [10]). The contribution of this paper is to extend the idea of optimizing revenue through capacity allocation to the context of telecommunications systems. (A simple treatment of optimizing profit through capacity allocation for a telecommunication system can be found in Jennings, Massey and McCalla [8] for example.) Specifically, we formulate a contract bidding process that applies to network systems, and develop mathematical machinery to carry out this process.

The paper is organized as follows. Section 2 presents the model formulation and outlines our basic approach. Section 3 describes the key mathematical techniques for solving the problem and formulates the calculation procedure. Numerical examples are demonstrated in Section 4, and concluding remarks are made in Section 5.

2. Formulation and solution approach.

2.1. Problem statement. Consider a carrier who has certain number of communication channels on a single link and operates in two markets:

> **a. Service market.** In the service market, customers start and end the use of a channel at random. The request for usage is either granted instantaneously upon their arrival, if available capacity exists, or blocked otherwise. The customer arrival process is assumed to be Poisson with a constant rate. Each connection lasts for a random period of time and all are assumed to be mutually independent and identically distributed. Moreover, if we assume that the number of channels used by these customers is in steady state, then ρ, the mean offered service load, equals the product of the customer arrival rate and the average duration of the connection. Suppose that the service market is highly competitive, and the carrier is the price taker. The service revenue per unit of time and per channel used is then given exogenously and denoted as r. Let β be the *blocking rate*, which is determined by the

mean offered service load ρ and the number of available channels. The expected revenue from the service market is then $(1-\beta)*r*\rho$.

b. Capacity market. In the capacity market, customers make advanced reservations for guaranteed use of part of the carrier's capacity for specific time periods. Suppose that the carrier sells capacity to those customers by signing a contract with them. Each contract allows a customer full usage of a channel from a specific starting time, T_b, to a specific ending time, T_e. Define $[T_b, T_e]$ as the *delivery period* and its duration $T_e - T_b$ as the *contract length.* Note that a customer can reserve multiple channels by signing multiple contracts with the carrier. All contracts are made between time 0 and the starting time of the delivery period T_b, where we define the period of $[0, T_b]$ as the *contracting window.*

During this window, customers arrive randomly according to a Poisson process with mean rate λ_c. Each arriving customer bids his/her willingness to pay for the contract, and the carrier can either accept or reject the bid. If a bid is accepted, the carrier is obliged to guarantee the customer the use of one channel throughout the delivery period, and the customer is obliged to pay the carrier the accepted bid price. As a convention of forward contracts, the carrier does not need to disclose the contracting information with a customer to a third party (see Chapter 1 of Musiela and Rutkowski [11]). Hence, the carrier's acceptance or rejection of a bid won't affect the arrival and bidding strategy of other customers.

Assume the joint distribution of bids is identical, independently distributed with a given probability density function $f(w)$. We also assume that both λ_c and $f(w)$ are known to the carrier. If not, the carrier can start with some prior estimates, and use a feedback-based estimation procedure, such as the one described in Keon and Anandlingam [7], to update these parameters.

In general, accepting a bid generates revenue that equals the bid price for the carrier, but reduces revenue from the service market as service customers are then served with less capacity. Therefore, the carrier should accept a bid only if the bid price can at least offset the loss of service revenue. We are interested in developing a policy for the carrier to make this acceptance decision. Specifically, let L_t $(0 \le t \le T_b)$ be the *number of channels* available at time t. Suppose a customer shows up at t and bids w for a contract. Should the carrier accept this bid?

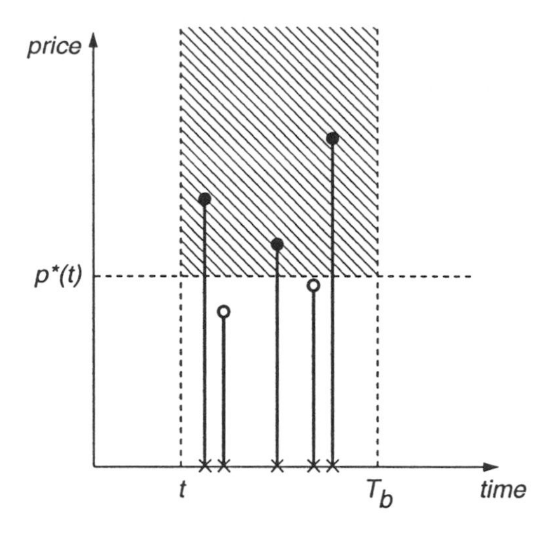

FIG. 1. *Arriving bids that are accepted and rejected at time t.*

2.2. Basic approach. One reasonable approach is to let the carrier set a *threshold price* $p(t)$, based on L_t, and accept the bid if and only if the bid price w exceeds $p(t)$. This is illustrated in Figure 1 where the $\times$'s along the time axis represent the arrival times of all bids. The heights of the vertical line segment above each $\times$ represents one realization of the random bid price each customer may give. All arriving bids during the time interval $[t, T_b]$ with a price that exceeds the threshold price $p(t) = p^*$ are represented here by a line segment that touches the shaded region. Such vertical line segments are topped by a black circle whose coordinates are the time of arrival and the bid price. Similar coordinates where the bid price does not exceed the threshold price are marked by an open circle. The number of black circles in the shaded region then correspond to the number of accepted bids.

The value of $p(t)$ is set to maximize the expected revenue from the capacity market during the remaining contracting period $[t, T_b]$, defined as *contracting revenue* and denoted as $E_t[\pi_c]$, and the expected revenue from the service market, defined as *service revenue* and denoted as $E_t[\pi_s]$, over the delivery period $[T_b, T_e]$.

Let M_t be the number of bids that arrive during the remaining contracting period $[t, T_b]$. If $E[M_t] = \lambda_c \cdot (T_b - t)$ is sufficiently small in comparison with the available capacity L_t, then:

$$E_t[\pi_c] = \sum_{m=0}^{L_t} Pr(M_t = m) \cdot E\left[\pi_c(t) \,|\, M_t = m\right]$$

$$
\begin{aligned}
&= \sum_{m=0}^{L_t} Pr(M_t = m) \cdot m \int_{p(t)}^{\infty} w f(w) dw \\
(2.1) \qquad &\approx E[M_t] \cdot \int_{p(t)}^{\infty} w f(w) dw \\
&= \lambda_c \cdot (T_b - t) \cdot \int_{p(t)}^{\infty} w f(w) dw.
\end{aligned}
$$

Given the threshold price $p(t)$, let ΔL_t be the number of bids that are accepted in period $[t, T_b]$. The process that counts the number of contract customer arrivals is Poisson and $\lambda_c \cdot (T_b - t)$ is sufficiently smaller than L_t, so the distribution of ΔL_t is approximately Poisson with a mean of

$$
(2.2) \qquad \Lambda_c(t) = \lambda_c \cdot (T_b - t) \cdot \int_{p(t)}^{\infty} f(w) dw.
$$

Since $L_b = L_t - \Delta L_t$ equals the number of channels that are available at T_b, and thus are used to serve end users, the revenue from the service market equals:

$$
\begin{aligned}
E_t[\pi_s] &= \sum_{l=0}^{L_t} Pr(\Delta L_t = l)\,(1 - \beta(L_b))\, r\rho \cdot (T_e - T_b) \\
(2.3) \qquad &= \sum_{l=0}^{L_t} \frac{\Lambda_c(t)^l}{l!} e^{-\Lambda_c(t)}\,(1 - \beta(L_t - l))\, r\,\rho \cdot (T_e - T_b) \\
&= \sum_{l=0}^{L_t} \frac{\Lambda_c(t)^l}{l!} e^{-\Lambda_c(t)} B_l,
\end{aligned}
$$

where

$$
(2.4) \qquad B_l = (1 - \beta(L_t - l))\, r\,\rho \cdot (T_e - T_b)
$$

and $\beta(\cdot)$ is the *Erlang blocking formula* (see Erlang [5]) as a function of the number of channels, with offered load ρ or

$$
(2.5) \qquad \beta(L) = \frac{\rho^L}{L!} \Big/ \sum_{l=0}^{L} \frac{\rho^l}{l!}\ .
$$

To summarize, the bid acceptance decision is now formulated as a problem of finding, for fixed t, a threshold price $p(t)$ to maximize the following expected revenue function:

$$
(2.6) \qquad E_t[\pi] = E_t[\pi_s] + E_t[\pi_c].
$$

2.3. Optimal threshold price. Below we use the convention that $1/(-1)! = 0$. For notational simplicity, we also write $p(t)$ and $\Lambda(t)$ as p and Λ respectively, suppressing their explicit time dependence. Taking derivatives of the expected contracting and service revenue functions with respect to p gives us:

$$\frac{dE_t[\pi_c]}{dp} = -\lambda_c \cdot (T_b - t)pf(p) < 0. \tag{2.7}$$

Given that $B_{L_t} = (1 - \beta(0))r\rho(T_e - T_b) = 0$, we also have

$$\frac{dE_t[\pi_s]}{dp} = \sum_{l=0}^{L_t} e^{-\Lambda_c} \left[\frac{\Lambda_c^{l-1}}{(l-1)!} - \frac{\Lambda_c^l}{l!}\right] B_l \frac{d\Lambda_c}{dp} \tag{2.8}$$

$$= -\lambda_c \cdot (T_b - t) f(p) \sum_{l=0}^{L_t - 1} e^{-\Lambda_c} \left[\frac{\Lambda_c^{l-1}}{(l-1)!} - \frac{\Lambda_c^l}{l!}\right] B_l \tag{2.9}$$

$$= \lambda_c \cdot (T_b - t) f(p) \sum_{l=0}^{L_t - 1} e^{-\Lambda_c} \frac{\Lambda_c^l}{l!} (B_l - B_{l+1}) \tag{2.10}$$

$$= \lambda_c \cdot (T_b - t) f(p) e^{-\Lambda_c} r \times \sum_{l=0}^{L_t - 1} \frac{\Lambda_c^l}{l!} (\beta(L_t - l - 1) - \beta(L_t - l))\, \rho \cdot (T_e - T_b) \tag{2.11}$$

$$> 0. \tag{2.12}$$

From (2.7) and (2.12), increasing p leads to an increase in the expected service revenue and a decrease in the expected contracting revenue. Therefore, the total expected revenue, $E_t[\pi_c] + E_t[\pi_s]$, is globally optimized at a unique point $p^*(t)$ that satisfies:

$$\frac{dE_t[\pi_c]}{dp} = -\frac{dE_t[\pi_s]}{dp}, \tag{2.13}$$

i.e.

$$p^*(t) = e^{-\Lambda_c(t)} r \cdot (T_e - T_b)\rho \times \sum_{l=0}^{L_t - 1} \frac{\Lambda_c(t)^l}{l!} (\beta(L_t - l - 1) - \beta(L_t - l)). \tag{2.14}$$

3. Algorithm. At each time t within the contracting window $[0, T_b]$, the carrier can determine the optimal threshold price, $p^*(t)$, from (2.14). Notice that $\Lambda_c(t)$ depends on $p^*(t)$, and appears on the right-hand side of the equation. Therefore, an iterative approach such as Newton's method needs to be applied to solve the equation. The critical step of the approach is to find an efficient technique to evaluate this sum of terms weighted by a Poisson distribution.

Let Q be a Poisson distribution with mean $\Lambda_c(t)$ and let $f(n) = \beta(L-n)$ be the function on the integers that vanishes outside the set $\{0, 1, \ldots, L\}$. We then have:

$$E[f(Q)] = \sum_{n=0}^{\infty} \frac{e^{-\Lambda_c(t)}\Lambda_c(t)^n}{n!} f(n) = \sum_{n=0}^{L} \frac{e^{-\Lambda_c(t)}\Lambda_c(t)^n}{n!} f(n). \tag{3.1}$$

If $N = \{ N(s) \mid s \geq 0 \}$ is a Poisson process with rate $\lambda = \Lambda_c(t)$, then

$$E\left[f\left(N(1)\right)\right] = E\left[f(Q)\right]. \tag{3.2}$$

Notice that $E\left[f\left(N(1)+1\right)\right] - E\left[f\left(N(1)\right)\right]$ is the same as the quantity we want to evaluate. Since N is a Markov process, we have for all f

$$\frac{d}{ds} E\left[f\left(N(s)\right)\right] = \lambda \left(E\left[f\left(N(s)+1\right)\right] - E\left[f\left(N(s)\right)\right]\right). \tag{3.3}$$

We now proceed to give a numerical scheme for computing $E\left[f\left(N(s)\right)\right]$ and $E\left[f\left(N(s)+1\right)\right]$.

Let $g_n(s) \equiv E\left[f\left(N(s)+n\right)\right]$. Using (3.3), we have

$$\frac{d}{ds} g_n(s) = \lambda\left(g_{n+1}(s) - g_n(s)\right). \tag{3.4}$$

Since f vanishes on all integers larger than $L_t - 1$, we have $g_{L+1} = 0$ where $L = L_t - 1$. In general, we can compute the vector $\mathbf{g}(s) = [g_0(s), \ldots, g_L(s)]$ by using a forward difference scheme to obtain

$$g_n(s + \Delta s) = (1 - \lambda \Delta s) g_n(s) + \lambda \Delta s g_{n+1}(s). \tag{3.5}$$

Note that:

$$g_n(0) = f(n) = \beta(L-n). \tag{3.6}$$

Based on (3.5) and (3.6), within a given time step, we start with $n = L$, decrease n down to 0, and obtain the desired result for the case of $s = 1$:

$$g_1(1) - g_0(1) = E\left[f\left(N(1)+1\right)\right] - E\left[f\left(N(1)\right)\right]. \tag{3.7}$$

The pseudo-code for this process is as follows:

```
time = 0.0;
g(L + 1) = 0.0;
for (n = L; n ≥ 0; n--){
        g(n) = f(n);
}
while ( time < 1 ){
        for (n = L; n ≥ 0; n--){
                g(n) *=  (1 − λ * Δs);
                g(n) +=  λ * Δs * g(n + 1);
        }
        time +=  Δs;
}
```

where we must keep $\Delta s \ll 1/\lambda$.

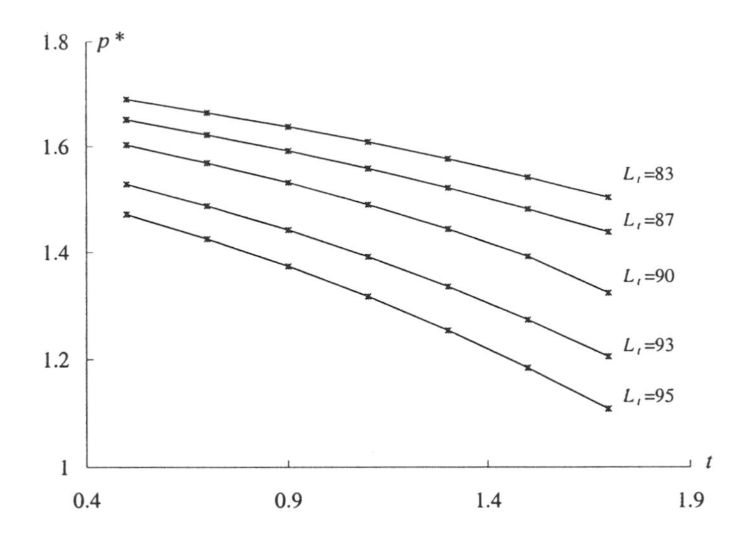

FIG. 2. *Plot of the optimal threshold price p^* vs. time t, given different numbers of available channels L_t.*

4. Numerical examples. In this section, we demonstrate the application of our approach through numerical examples. We start with a base case scenario in which the offered service load $\rho = 90$ and the reward for serving end users per used channel per unit of time is $r = 1$. Assume that in the capacity market, each contract grants the use of one channel within the period $[T_b, T_e]$ where $T_b = 2$ and $T_e = 4$, so the contract length is $T_e - T_b = 2$. Within the contracting window $[0, T_b]$, bids for a contract arrive according to a Poisson process with mean rate, $\lambda_c = 10$. The bids are exponentially distributed with a mean of 0.2, i.e. $f(w) = 0.2e^{-0.2w}$.

Let L_t be the number of channels available at some time $t < T_b$, where $[t, T_b]$ is the remaining period to make capacity contracts. Figure 2 shows that given the same number of available channels, the larger the value of t (i.e. a smaller amount of time remains to make contracts), the lower the optimal threshold price, p^*. Nevertheless, L_t usually decreases as t increases because more contracts are made during a longer period. The figure shows that p^* can be higher at a later t if L_t is smaller. Therefore, the optimal threshold price may decrease or increase over time.

We now vary some parameters, and discuss the behavior of the optimal threshold price at a given time $t = 1$.

4.1. Effects of service load. Figure 3 shows the optimal threshold price (p^*) at $t = 1$, given different levels of offered service load, ρ. With higher value of ρ, each channel generates more service revenues, so it is no surprise that p^* increases with ρ.

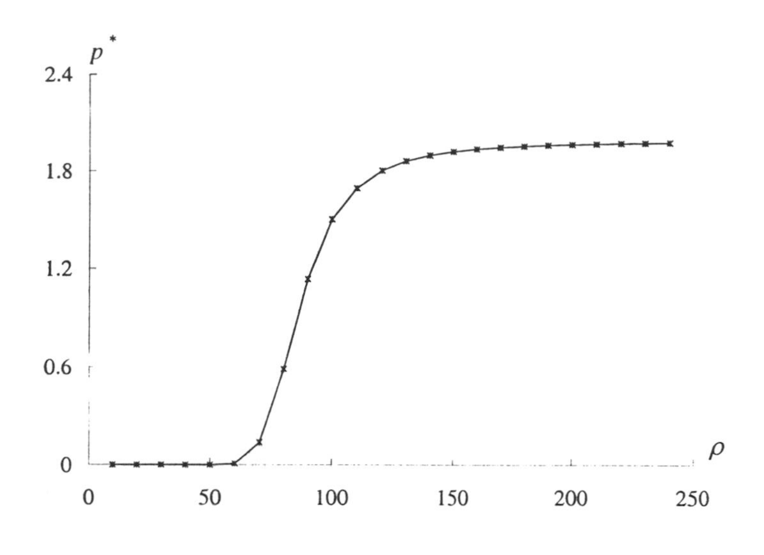

FIG. 3. *Plot of the offered load ρ vs. the optimal threshold price p^*.*

It is interesting to see that p^* flattens at both ends of ρ. When the network is lightly used by service customers, selling capacity to contract customers does not reduce service revenue significantly. Therefore, the carrier should set p^* to zero and accept all positive bids for capacities. This explains why in Figure 3, there exists a range of low values for ρ where p^* stays close to zero. If the network is heavily loaded with service customers, each channel is used almost all the time. In those situations, selling a channel to contract customers is profitable if and only if the price of the contract exceeds $r(T_e - T_b)$, i.e. the contract length multiplied by the service revenue per channel per unit of time. Hence, $r \cdot (T_e - T_b)$ is the upper limit of p^*, and is reached when ρ becomes sufficiently high.

When the network is neither under-loaded nor overloaded, p^* is sensitive to ρ. As can be seen from the figure, there exists a range of ρ values, which we define as the critical region, where the optimal threshold price rises rapidly from near zero to its upper-limit. Our approach is the most useful in that region where p^* can be quite different with respect to small changes in ρ, and is close to neither zero, nor the upper-limit.

Starting and ending points of the critical region depend on the arrival rate of contracting customers, λ_c. As shown in Figure 4, with a higher value of λ_c, the critical region starts and ends at lower levels of ρ.

4.2. Optimal threshold price and contract length. We now consider how the optimal threshold price p^* varies with contract length $T_e - T_b$. Intuitively, one might expect that everything else being equal, if the contract length is k times as long as the base case, then a contract costs k times

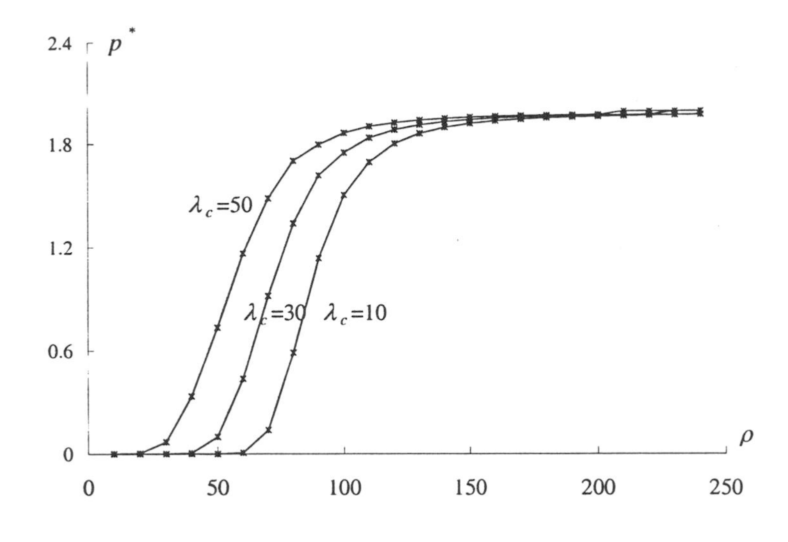

FIG. 4. *Plot of the offered load ρ vs. the optimal threshold price p^*, with different arrival rates for contracting customers λ_c.*

as much as service revenue. Therefore, the optimal threshold price should be proportional to the contract length. However, our numerical examples show that the conjecture is only true when the network is overly loaded.

Table 1 shows p^* at $t = 1$ given different values of contract length $T_e - T_b$ and service load ρ. In the base case, we assume $T_e - T_b = 1$, and show values of p^*, given $\rho = 80$, 90, 120, 150 (assume the number of available channels, $L_t = 100$). We then increase $T_e - T_b$ from 1 to 8, and show the percentage change of p^*. It can be seen from the table that when $\rho = 80$ or 90 (i.e. the service load is 80% or 90% of the available capacity), then the percentage increase of p^* is slower than the percentage increase of $T_e - T_b$. When the ρ reaches 120, 150, p^* increases linearly with $T_e - T_b$.

The explanation of this result lies on the fact that increasing p^* reduces the number of contracts to be accepted, and thus leaves more channels to the service customers. When the network is not over-loaded, increasing the number of channels to accommodate a given service load reduces the channel usage rate. Therefore, the per channel service revenue per unit of time decreases. This means that the cost of selling channels by contract does not increase in proportion to the contract length if the threshold price changes. As a result, p^* also does not increase proportionally to the contract length. However, if the network is sufficiently overloaded with service customers, the channel usage rate stays at the maximum regardless how many contracts have been accepted. In this case, per channel service revenue per unit of time is insensitive to a change in the threshold price. It

TABLE 1
Percentage increase of the optimal threshold price with the contract length.

	$\rho = 80$	$\rho = 90$	$\rho = 120$	$\rho = 200$
base case p^* $T_e - T_b = 1$	0.310	0.590	0.902	0.984
$T_e - T_b$				
2	190%	193%	200%	200%
3	272%	278%	300%	300%
4	347%	362%	400%	400%
5	418%	434%	493%	500%
6	485%	511%	590%	600%
7	548%	582%	686%	700%
8	614%	652%	781%	800%

follows that the cost of selling a channel by contract, and thus p^* increases linearly with the contract length.

The above result suggests that in a normally loaded network, a discount in terms of the price per channel per unit of time, should be given to contracts with longer duration.

5. Conclusions. In this paper, we study a mechanism for a communications carrier to sell its network capacity through a forward contract. We develop a threshold pricing approach where the carrier accepts a bid for the contract only when the bidder is willing to pay a price that at least offsets the loss of service revenue due to a smaller number of channels. The approach maximizes the carrier's total expected revenue from both service and capacity markets. We formulate a feedback-based process for determining the optimal threshold price, and derive key mathematical techniques for price calculation. Under our approach, the optimal price may increase or decrease over time, is close to zero in a lightly loaded network, and reaches its upper-bound in a heavily-loaded network. We also find it is optimal for a carrier to discount contracts with longer duration, unless the network is overloaded.

Our work can be extended in many directions. We can directly apply the analysis of this paper to the case of time inhomogeneous arrivals. We can model M_t as a time inhomogeneous Poisson process so that all increments $M_s - M_t$ for $s < t$ have a Poisson distribution and

$$E[M_s - M_t] = \int_s^t \lambda_c(\tau)\, d\tau. \tag{5.1}$$

Moreover, the resulting ΔL_t still has a Poisson distribution since it is the "thinning" of the Poisson process M_t (see Daley and Vere Jones [2] for details) where we now have

$$\Lambda_c(t) = \int_t^{T_b} \lambda_c(\tau) \int_{p(\tau)}^{\infty} f(w) dw \, d\tau. \tag{5.2}$$

Using the *modified offered load approximation*, see Massey and Whitt [9], we can approximate the blocking rate for the service market by still using the Erlang blocking formula but at time t we set the new offered load equal to

$$\rho = E\left[\int_{t-S}^{t} \lambda_s(\tau)\, d\tau\right], \tag{5.3}$$

where λ_s is the mean arrival rate function for the customers in the service market and S is a random connection holding time. We assume that the holding times of all customers are independent and identically distributed.

Time dependent arrival phenomena, such as a surge in customer traffic at a specific time, can be approximated well by these approximate methods for times of low blocking (under 10%). One effect that arises here is the blocking rate is no longer insensitive to the distribution of the holding time. In Davis, Massey and Whitt [3] we demonstrate this phenomenon for the offered service load. Quantities of interest like the lag between the times of peak demand and the times of peak load will depend on more than the first moments of the holding time. So for a fixed mean holding time, a distribution with a more heavy tailed distribution has a longer lag time. For a discussion of the behavior for the offered service load, see Eick, Massey and Whitt [4].

More general techniques need to be developed to apply our approach to networks with multiple links and multiple-classes of services. It is also interesting to examine a carrier who is not only interested in selling but also buying capacity through forward contracts. Furthermore, we can also look into situations in which a contract can be resold to other bidders, as is the case in the current energy capacity market.

REFERENCES

[1] Courcoubetis, C., Dimakis, A. and Reiman, M.I. *Providing bandwidth guarantees over a best-effort network: call-admission and pricing,* Proceedings of IEEE INFOCOM 2001, pp. 459–467, April.

[2] Daley, D.J. and Vere-Jones, D. *An Introduction to the Theory of Point Processes*, Springer, New York, 1988.

[3] Davis, J.L., Massey, W.A. and Whitt, W. *Sensitivity to the Service-Time Distribution in the Nonstationary Erlang Loss Model*, Management Science, 41:6 (June 1995), pp. 1107–1116.

[4] Eick, S., Massey, W.A. and Whitt, W. *The Physics of the $M_t/G/\infty$ Queue*, Operations Research, 41 (1993) 731–742.

[5] A.K. Erlang, *Solutions of Some Problems in the Theory of Probabilities of Significance in Automatic Telephone Exchanges*, The Post Office Electrical Engineers' Journal (Translated from the 1917 article in Danish in *Elektroteknikeren* vol. 13), **10** (1918) 189–197.

[6] JENNINGS, O.B., MASSEY, W.A. AND MCCALLA, C. *Optimal Profit for Leased Lines Services*, Proceedings of the 15th International Teletraffic Congress - ITC 15 (editors V. Ramaswami and P. E. Wirth), June 1997, pp. 803–814.

[7] KEON, N. AND ANANDLINGAM, G. *Real Time Pricing of Multiple Telecommunications Services under Uncertain Demand*, Proceedings of 7th International Conference on Telecommunications Systems, Modeling and Analysis, pp. 28–47, March 1999.

[8] LANNING, S.G, MASSEY, W.A., RIDER, B. AND WANG, Q. *Optimal Pricing in Queueing Systems with Quality of Service Constraints*, Proceedings of the 16th International Teletraffic Congress - ITC 16, June 1999, pp. 747–756.

[9] MASSEY, W.A. AND WHITT W. *On the Modified-Offered-Load Approximation for the Nonstationary Erlang Loss Model*, The Fundamental Role of Teletraffic in the Evolution of Telecommunication Networks (14th International Teletraffic Congress). June 1994, pp. 145–153.

[10] MCGILL, J.L. AND VAN RYZIN, G.J., *Revenue Management: Research Overview and Prospects*, Transportation Science 33, 233–256 (1999).

[11] MUSIELA, M. AND RUTKOWSKI, M. *Martingale Methods in Financial Modelling.* Springer Verlag, New York, 1997.

[12] WISCHIK, D. AND GREENBERG, A. *Admission Control for Booking Ahead Shared Resources*, IEEE InfoCom, 1998.

LIST OF "HOT TOPICS" WORKSHOP PARTICIPANTS

- Philipp Afeche, MEDS, Northwestern University
- Beth Allen, Department of Economics, University of Minnesota
- G. Anandalingam, R.H. Smith School of Business, University of Maryland
- Kevin Anderson, Institute for Mathematics and its Applications
- Damian Beil, Operations Research Center, Massachusetts Institute of Technology
- Saif Benjaafar, Department of Mechanical Engineering, University of Minnesota
- Santiago Betelu, Institute for Mathematics and its Applications
- Sushil Bikhchandani, Anderson Graduate School of Management, University of California Los Angeles
- John Birge, School of Engineering and Applied Sciences, Northwestern University
- Shantanu Biswas, Computer Science and Automation, Indian Institute of Science
- Chris Caplice, Logistics.com
- Jamylle Carter, Institute for Mathematics and its Applications
- James Case
- Li-Tien Cheng, Institute for Mathematics and its Applications
- John Collins, Department of Computer Science, University of Minnesota
- William L. Cooper, Department of Mechanical Engineering, University of Minnesota
- Janos A. Csirik, Mathematics and Cryptography, AT&T Labs - Research
- Brenda Dietrich, Mathematical Sciences Department, IBM T.J. Watson Research Center
- Fred Dulles, Institute for Mathematics and its Applications
- Michael Ekhaus, Gibraltar Analytics
- Wedad Elmaghraby, School of Industrial and Systems Engineering, Georgia Institute of Technology
- Selim Esedoglu, Institute for Mathematics and its Applications
- Marta Eso, IBM Research Division, Thomas J. Watson Research Center
- Peter Eso, MEDS, Kellogg Graduate School of Management, Northwestern University
- Liping Fang, School of Industrial Engineering, Ryerson Polytechnic University
- Jeremie Gallien, Neoptis, Inc.

- Maria Gini, Department of Computer Science and Engineering, University of Minnesota
- Jeremy Goecks, Computer Science, University of Minnesota
- Kemal Guler, Decision Technologies Department, Hewlett Packard Laboratories
- Christiaan Hogendorn, Technology Management and Economics Research, Lucent Technologies-Bell Laboratories
- Suzhou Huang, Ford Research Laboratory, Ford Motor Company
- Shailendra Jain, Decision Technology Department, Hewlett-Packard Laboratories
- Jayant R. Kalagnanam, IBM T.J. Watson Research Center
- Howard Karloff, Research Department, AT&T Labs
- Elena Katok, Management Science and Information Systems, Pennsylvania State University
- Neil Keon, Edwin L. Cox School of Business, Southern Methodist University
- Jeff Kephart, IBM T.J. Watson Research Center
- Pinar Keskinocak, School of Industrial and Systems Engineering, Georgia Institute of Technology
- Sanjeev Khudanpur, Center for Language and Speech Processing, Johns Hopkins University
- John Kickhaut, Carlson School of Management, University of Minnesota
- Erhan Kutanoglu, Department of Industrial Engineering, University of Arkansas
- Han La Poutre, Computer Science Department, CWI
- Steven G. Lanning, Aerie Networks
- John Ledyard, Division of Humanities and Social Sciences, California Institute of Technology
- Meg Ledyard, Economics Department, University of Minnesota
- Jenny Li, Mathematic and Economics, Penn State University
- Costis Maglaras, Graduate School of Business, Columbia University
- Andrew Mclennan, Department of Economics, University of Minnesota
- Paul Milgrom, Department of Economics, Stanford University
- Willard Miller, Institute for Mathematics and its Applications
- Clyde Monma, Information and Computer Sciences, Telcordia Technologies
- Karl Nilsson, University of Minnesota
- Andrew Odlyzko, Mathematics and Cryptogoraphy Research, AT&T Labs - Research
- David Parkes, Computer and Information Science, University of Pennsylvania

- S. Raghavan, The R.H. Smith School of Business, University of Maryland, College Park
- Alvin Roth, Department of Economics, Harvard University
- Michael H. Rothkopf, RUTCOR and Faculty of Management, Rutgers University
- Irv Salmeen, Ford Motor Company
- Kameshwaran Sampath, Computer Science and Automation, Indian Institute of Science
- James Schummer, Kellogg Graduate School of Management, Northwestern University
- Stephen Simpson, Department of Mathematics, Penn State University
- Kerem Tomak, Management Science and Information Systems, University of Texas at Austin
- Garrett van Ryzin, School of Business, Columbia University
- Arun Verma, Cornell Theory Center, Cornell University
- Rakesh V. Vohra, J.L. Kellogg Graduate School of Management, Northwestern University
- Gustavo Vulcano, MS/OM, Columbia Business School
- Lan Wang, Science Research Lab, Ford Motor Company
- Robert Weber, Kellogg Graduate School of Management, Northwestern University
- Larry Wein, Sloan School of Management, Massachusetts Institute of Technology
- Michael Wellman, Artificial Intelligence Laboratory, University of Michigan
- Yong Yang, Ford Research Laborator, Ford Motor Company

IMA SUMMER PROGRAMS

1987	Robotics
1988	Signal Processing
1989	Robust Statistics and Diagnostics
1990	Radar and Sonar (June 18–29)
	New Directions in Time Series Analysis (July 2–27)
1991	Semiconductors
1992	Environmental Studies: Mathematical, Computational, and Statistical Analysis
1993	Modeling, Mesh Generation, and Adaptive Numerical Methods for Partial Differential Equations
1994	Molecular Biology
1995	Large Scale Optimizations with Applications to Inverse Problems, Optimal Control and Design, and Molecular and Structural Optimization
1996	Emerging Applications of Number Theory (July 15–26)
	Theory of Random Sets (August 22–24)
1997	Statistics in the Health Sciences
1998	Coding and Cryptography (July 6–18)
	Mathematical Modeling in Industry (July 22–31)
1999	Codes, Systems, and Graphical Models (August 2–13, 1999)
2000	Mathematical Modeling in Industry: A Workshop for Graduate Students (July 19–28)
2001	Geometric Methods in Inverse Problems and PDE Control (July 16–27)
2002	Special Functions in the Digital Age (July 22–August 2)

IMA "HOT TOPICS" WORKSHOPS

- Challenges and Opportunities in Genomics: Production, Storage, Mining and Use, April 24–27, 1999
- Decision Making Under Uncertainty: Energy and Environmental Models, July 20–24, 1999
- Analysis and Modeling of Optical Devices, September 9–10, 1999
- Decision Making under Uncertainty: Assessment of the Reliability of Mathematical Models, September 16–17, 1999
- Scaling Phenomena in Communication Networks, October 22–24, 1999
- Text Mining, April 17–18, 2000
- Mathematical Challenges in Global Positioning Systems (GPS), August 16–18, 2000
- Modeling and Analysis of Noise in Integrated Circuits and Systems, August 29–30, 2000
- Mathematics of the Internet: E-Auction and Markets, December 3–5, 2000
- Analysis and Modeling of Industrial Jetting Processes, January 10–13, 2001

- Special Workshop: Mathematical Opportunities in Large-Scale Network Dynamics, August 6–7, 2001
- Wireless Networks, August 8–10 2001

SPRINGER LECTURE NOTES FROM THE IMA:

The Mathematics and Physics of Disordered Media
Editors: Barry Hughes and Barry Ninham
(Lecture Notes in Math., Volume 1035, 1983)

Orienting Polymers
Editor: J.L. Ericksen
(Lecture Notes in Math., Volume 1063, 1984)

New Perspectives in Thermodynamics
Editor: James Serrin
(Springer-Verlag, 1986)

Models of Economic Dynamics
Editor: Hugo Sonnenschein
(Lecture Notes in Econ., Volume 264, 1986)

The IMA Volumes in Mathematics and its Applications

Current Volumes:

23 **Signal Processing Part II: Control Theory and Applications of Signal Processing** L. Auslander, F.A. Grünbaum, J.W. Helton, T. Kailath, P. Khargonekar, and S. Mitter (eds.)

24 **Mathematics in Industrial Problems, Part 2** A. Friedman

25 **Solitons in Physics, Mathematics, and Nonlinear Optics** P.J. Olver and D.H. Sattinger (eds.)

26 **Two Phase Flows and Waves** D.D. Joseph and D.G. Schaeffer (eds.)

27 **Nonlinear Evolution Equations that Change Type** B.L. Keyfitz and M. Shearer (eds.)

28 **Computer Aided Proofs in Analysis** K. Meyer and D. Schmidt (eds.)

29 **Multidimensional Hyperbolic Problems and Computations** A. Majda and J. Glimm (eds.)

30 **Microlocal Analysis and Nonlinear Waves** M. Beals, R. Melrose, and J. Rauch (eds.)

31 **Mathematics in Industrial Problems, Part 3** A. Friedman

32 **Radar and Sonar, Part I** R. Blahut, W. Miller, Jr., and C. Wilcox

33 **Directions in Robust Statistics and Diagnostics: Part I** W.A. Stahel and S. Weisberg (eds.)

34 **Directions in Robust Statistics and Diagnostics: Part II** W.A. Stahel and S. Weisberg (eds.)

35 **Dynamical Issues in Combustion Theory** P. Fife, A. Liñán, and F.A. Williams (eds.)

36 **Computing and Graphics in Statistics** A. Buja and P. Tukey (eds.)

37 **Patterns and Dynamics in Reactive Media** H. Swinney, G. Aris, and D. Aronson (eds.)

38 **Mathematics in Industrial Problems, Part 4** A. Friedman

39 **Radar and Sonar, Part II** F.A. Grünbaum, M. Bernfeld, and R.E. Blahut (eds.)

40 **Nonlinear Phenomena in Atmospheric and Oceanic Sciences** G.F. Carnevale and R.T. Pierrehumbert (eds.)

41 **Chaotic Processes in the Geological Sciences** D.A. Yuen (ed.)

42 **Partial Differential Equations with Minimal Smoothness and Applications** B. Dahlberg, E. Fabes, R. Fefferman, D. Jerison, C. Kenig, and J. Pipher (eds.)

43 **On the Evolution of Phase Boundaries** M.E. Gurtin and G.B. McFadden

44 **Twist Mappings and Their Applications** R. McGehee and K.R. Meyer (eds.)

45 **New Directions in Time Series Analysis, Part I**
D. Brillinger, P. Caines, J. Geweke, E. Parzen, M. Rosenblatt, and M.S. Taqqu (eds.)

46 **New Directions in Time Series Analysis, Part II**
D. Brillinger, P. Caines, J. Geweke, E. Parzen, M. Rosenblatt, and M.S. Taqqu (eds.)

47 **Degenerate Diffusions**
W.-M. Ni, L.A. Peletier, and J.-L. Vazquez (eds.)

48 **Linear Algebra, Markov Chains, and Queueing Models**
C.D. Meyer and R.J. Plemmons (eds.)

49 **Mathematics in Industrial Problems, Part 5** A. Friedman

50 **Combinatorial and Graph-Theoretic Problems in Linear Algebra**
R.A. Brualdi, S. Friedland, and V. Klee (eds.)

51 **Statistical Thermodynamics and Differential Geometry of Microstructured Materials**
H.T. Davis and J.C.C. Nitsche (eds.)

52 **Shock Induced Transitions and Phase Structures in General Media** J.E. Dunn, R. Fosdick, and M. Slemrod (eds.)

53 **Variational and Free Boundary Problems**
A. Friedman and J. Spruck (eds.)

54 **Microstructure and Phase Transitions**
D. Kinderlehrer, R. James, M. Luskin, and J.L. Ericksen (eds.)

55 **Turbulence in Fluid Flows: A Dynamical Systems Approach**
G.R. Sell, C. Foias, and R. Temam (eds.)

56 **Graph Theory and Sparse Matrix Computation**
A. George, J.R. Gilbert, and J.W.H. Liu (eds.)

57 **Mathematics in Industrial Problems, Part 6** A. Friedman

58 **Semiconductors, Part I**
W.M. Coughran, Jr., J. Cole, P. Lloyd, and J. White (eds.)

59 **Semiconductors, Part II**
W.M. Coughran, Jr., J. Cole, P. Lloyd, and J. White (eds.)

60 **Recent Advances in Iterative Methods**
G. Golub, A. Greenbaum, and M. Luskin (eds.)

61 **Free Boundaries in Viscous Flows**
R.A. Brown and S.H. Davis (eds.)

62 **Linear Algebra for Control Theory**
P. Van Dooren and B. Wyman (eds.)

63 **Hamiltonian Dynamical Systems: History, Theory, and Applications**
H.S. Dumas, K.R. Meyer, and D.S. Schmidt (eds.)

64 **Systems and Control Theory for Power Systems**
J.H. Chow, P.V. Kokotovic, R.J. Thomas (eds.)

65 **Mathematical Finance**
M.H.A. Davis, D. Duffie, W.H. Fleming, and S.E. Shreve (eds.)

66 **Robust Control Theory** B.A. Francis and P.P. Khargonekar (eds.)
67 **Mathematics in Industrial Problems, Part 7** A. Friedman
68 **Flow Control** M.D. Gunzburger (ed.)
69 **Linear Algebra for Signal Processing**
A. Bojanczyk and G. Cybenko (eds.)
70 **Control and Optimal Design of Distributed Parameter Systems**
J.E. Lagnese, D.L. Russell, and L.W. White (eds.)
71 **Stochastic Networks** F.P. Kelly and R.J. Williams (eds.)
72 **Discrete Probability and Algorithms**
D. Aldous, P. Diaconis, J. Spencer, and J.M. Steele (eds.)
73 **Discrete Event Systems, Manufacturing Systems, and Communication Networks**
P.R. Kumar and P.P. Varaiya (eds.)
74 **Adaptive Control, Filtering, and Signal Processing**
K.J. Åström, G.C. Goodwin, and P.R. Kumar (eds.)
75 **Modeling, Mesh Generation, and Adaptive Numerical Methods for Partial Differential Equations** I. Babuska, J.E. Flaherty, W.D. Henshaw, J.E. Hopcroft, J.E. Oliger, and T. Tezduyar (eds.)
76 **Random Discrete Structures** D. Aldous and R. Pemantle (eds.)
77 **Nonlinear Stochastic PDEs: Hydrodynamic Limit and Burgers' Turbulence** T. Funaki and W.A. Woyczynski (eds.)
78 **Nonsmooth Analysis and Geometric Methods in Deterministic Optimal Control** B.S. Mordukhovich and H.J. Sussmann (eds.)
79 **Environmental Studies: Mathematical, Computational, and Statistical Analysis** M.F. Wheeler (ed.)
80 **Image Models (and their Speech Model Cousins)**
S.E. Levinson and L. Shepp (eds.)
81 **Genetic Mapping and DNA Sequencing**
T. Speed and M.S. Waterman (eds.)
82 **Mathematical Approaches to Biomolecular Structure and Dynamics**
J.P. Mesirov, K. Schulten, and D. Sumners (eds.)
83 **Mathematics in Industrial Problems, Part 8** A. Friedman
84 **Classical and Modern Branching Processes**
K.B. Athreya and P. Jagers (eds.)
85 **Stochastic Models in Geosystems**
S.A. Molchanov and W.A. Woyczynski (eds.)
86 **Computational Wave Propagation**
B. Engquist and G.A. Kriegsmann (eds.)
87 **Progress in Population Genetics and Human Evolution**
P. Donnelly and S. Tavaré (eds.)
88 **Mathematics in Industrial Problems, Part 9** A. Friedman
89 **Multiparticle Quantum Scattering With Applications to Nuclear, Atomic and Molecular Physics** D.G. Truhlar and B. Simon (eds.)

90 **Inverse Problems in Wave Propagation** G. Chavent, G. Papanicolau, P. Sacks, and W.W. Symes (eds.)
91 **Singularities and Oscillations** J. Rauch and M. Taylor (eds.)
92 **Large-Scale Optimization with Applications, Part I: Optimization in Inverse Problems and Design** L.T. Biegler, T.F. Coleman, A.R. Conn, and F. Santosa (eds.)
93 **Large-Scale Optimization with Applications, Part II: Optimal Design and Control** L.T. Biegler, T.F. Coleman, A.R. Conn, and F. Santosa (eds.)
94 **Large-Scale Optimization with Applications, Part III: Molecular Structure and Optimization** L.T. Biegler, T.F. Coleman, A.R. Conn, and F. Santosa (eds.)
95 **Quasiclassical Methods** J. Rauch and B. Simon (eds.)
96 **Wave Propagation in Complex Media** G. Papanicolaou (ed.)
97 **Random Sets: Theory and Applications** J. Goutsias, R.P.S. Mahler, and H.T. Nguyen (eds.)
98 **Particulate Flows: Processing and Rheology** D.A. Drew, D.D. Joseph, and S.L. Passman (eds.)
99 **Mathematics of Multiscale Materials** K.M. Golden, G.R. Grimmett, R.D. James, G.W. Milton, and P.N. Sen (eds.)
100 **Mathematics in Industrial Problems, Part 10** A. Friedman
101 **Nonlinear Optical Materials** J.V. Moloney (ed.)
102 **Numerical Methods for Polymeric Systems** S.G. Whittington (ed.)
103 **Topology and Geometry in Polymer Science** S.G. Whittington, D. Sumners, and T. Lodge (eds.)
104 **Essays on Mathematical Robotics** J. Baillieul, S.S. Sastry, and H.J. Sussmann (eds.)
105 **Algorithms For Parallel Processing** M.T. Heath, A. Ranade, and R.S. Schreiber (eds.)
106 **Parallel Processing of Discrete Problems** P.M. Pardalos (ed.)
107 **The Mathematics of Information Coding, Extraction, and Distribution** G. Cybenko, D.P. O'Leary, and J. Rissanen (eds.)
108 **Rational Drug Design** D.G. Truhlar, W. Howe, A.J. Hopfinger, J. Blaney, and R.A. Dammkoehler (eds.)
109 **Emerging Applications of Number Theory** D.A. Hejhal, J. Friedman, M.C. Gutzwiller, and A.M. Odlyzko (eds.)
110 **Computational Radiology and Imaging: Therapy and Diagnostics** C. Börgers and F. Natterer (eds.)
111 **Evolutionary Algorithms** L.D. Davis, K. De Jong, M.D. Vose, and L.D. Whitley (eds.)
112 **Statistics in Genetics** M.E. Halloran and S. Geisser (eds.)

113 **Grid Generation and Adaptive Algorithms** M.W. Bern, J.E. Flaherty, and M. Luskin (eds.)

114 **Diagnosis and Prediction** S. Geisser (ed.)

115 **Pattern Formation in Continuous and Coupled Systems: A Survey Volume** M. Golubitsky, D. Luss, and S.H. Strogatz (eds.)

116 **Statistical Models in Epidemiology, the Environment, and Clinical Trials** M.E. Halloran and D. Berry (eds.)

117 **Structured Adaptive Mesh Refinement (SAMR) Grid Methods** S.B. Baden, N.P. Chrisochoides, D.B. Gannon, and M.L. Norman (eds.)

118 **Dynamics of Algorithms** R. de la Llave, L.R. Petzold, and J. Lorenz (eds.)

119 **Numerical Methods for Bifurcation Problems and Large-Scale Dynamical Systems** E. Doedel and L.S. Tuckerman (eds.)

120 **Parallel Solution of Partial Differential Equations** P. Bjørstad and M. Luskin (eds.)

121 **Mathematical Models for Biological Pattern Formation** P.K. Maini and H.G. Othmer (eds.)

122 **Multiple-Time-Scale Dynamical Systems** C.K.R.T. Jones and A. Khibnik (eds.)

123 **Codes, Systems, and Graphical Models** B. Marcus and J. Rosenthal (eds.)

124 **Computational Modeling in Biological Fluid Dynamics** L.J. Fauci and S. Gueron (eds.)

125 **Mathematical Approaches for Emerging and Reemerging Infectious Diseases: An Introduction** C. Castillo-Chavez with S. Blower, P. van den Driessche, D. Kirschner, and A.A. Yakubu (eds.)

126 **Mathematical Approaches for Emerging and Reemerging Infectious Diseases: Models, Methods, and Theory** C. Castillo-Chavez with S. Blower, P. van den Driessche, D. Kirschner, and A.A. Yakubu (eds.)

127 **Mathematics of the Internet: E-Auction and Markets** B. Dietrich and R.V. Vohra (eds.)